# Letters to a Young Chemist

# Letters to a Young Chemist

Edited by

**Abhik Ghosh**
University of Tromsø, Norway

*A John Wiley & Sons, Inc., Publication*

For general information on our other products and services or for technical support, please
contact our Customer Care Department within the United States at (800) 762-2974, outside the
United States at (317) 572-3993 or fax (317) 572-4002.

Wiley also publishes its books in a variety of electronic formats. Some content that appears in
print may not be available in electronic formats. For more information about Wiley products,
visit our web site at www.wiley.com.

*Library of Congress Cataloging-in-Publication Data:*

Letters to a young chemist / edited by Abhik Ghosh.
    p. cm.
  Includes index.
  ISBN 978-0-470-39043-6 (pbk.)
  1. Chemistry--Vocational guidance.   2. Chemistry--Research. 3. Chemists--Biography.
I. Ghosh, Abhik.
  QD39.5.L58 2011
  540.23--dc22

                            2010039501

10  9  8  7  6  5  4  3  2

# Contents

# Foreword

**Stephen J. Lippard**

*Department of Chemistry, Massachusetts Institute
of Technology, Cambridge, MA 02139, USA*

This book should be required reading for all faculty members who teach chemistry at the high school, college, and university levels. Addressed primarily to college students, it fills an important gap by offering in conversational language a view into the contributions of chemistry and of chemists who are improving the human condition. Each of the chapters is written in the form of a letter to "Angela," a hypothetical undergraduate at UCSD, indicating exciting challenges in the fields represented by the authors, all of whom are chemists. Four broad areas are targeted for coverage, namely, applications of chemistry fundamentals, chemistry and the life sciences, functional materials, and chemistry and energy. The styles of the individual contributions are largely informal but vary somewhat across the 17 contributions, reflecting the taste of the individual authors. In order to give the reader a flavor for the offerings of this book, I select here examples to illustrate how some of the authors approached the challenge of stimulating a young mind to appreciate the excitement that chemists find in their subject. But first I offer a few introductory comments based on my own experience as an educator and researcher over more than four decades.

The subject of chemistry may be characterized by the size of the entities it investigates, falling between physics, which studies fundamental particles, and biology, which focuses on macromolecules, cells, and whole organisms. Some therefore refer to chemistry as a "central science," but this moniker is one that I personally find inadequate. In many respects, *synthesis* is the heart of chemistry. In thinking about the synthesis of new substances, there is nothing "central" about what chemists do, nor is it our mantra to serve our sister fields, although our synthetic constructs often do so both intentionally and serendipitously. Our core is to understand the making and breaking of chemical bonds, and in so doing we are able to devise preparative routes to novel molecules and supramolecular constructs as well as solids that are nowadays referred to as nanoparticles. Through synthesis, chemists provide society with new materials that can transform the way in which we go about our daily tasks. Along the way, we apply indirect methods to construct motion picture images of chemical reactions that, for the most part, occur on a spatial or temporal scale too small or too rapid to be visualized. Theoretical contributions can provide insights to support and extend experimental findings. The discoveries are thrilling and deeply satisfying for the knowledge they provide, but to the extent that they also serve society they justify the considerable expenditure of public funds required to build or purchase the reagents, apparatus, and facilities required to carry out chemical research. This service is described in considerable detail in the letters to Angela contained in this book and can also be appreciated through the educational contributions of chemists who share their knowledge of chemistry with students preparing for careers in the related fields of medicine, engineering, materials science, biology, and even patent law, to name but a few.

*Letters to a Young Chemist* offers significant ammunition for motivating young Angela to consider chemistry as a career. The book displays a variety of personal accounts written by a collection of chemistry faculty representing nearly every branch of the discipline, a full spectrum of ages and experience, and good female gender representation. The lack of participation by faculty at principally undergraduate institutions may perhaps be excused by the focus on research. There are many inspirational passages. Prof. Carl Wamser of Portland State University, assuming the persona of Angela's "Uncle Carl," one of many in a large family of chemists that includes his 97-year-old father and who still reads the chemical journals, writes his fictional niece that "Clean Electrons and Molecules Will Save the World." The topic is energy and the focus is on solar conversion, a highly popular subject among students today. Uncle Carl, reprising a tactic taken by the late Rick Smalley in addressing his audiences during speaking engagements, asks Angela and the reader to list the top ten issues that need to be addressed to improve the quality of life by the middle of the twenty-first century. Energy is one of them of course, which provides the lead-in to the rest of Carl's letter, but the others are worth repeating here because, in many instances, one can make a case that chemistry will be required for success. They are, in alphabetical order:

Democracy
Disease
Education
Energy
Environment
Food
Population
Poverty
Terrorism/war
Water

Disease may be interpreted to include the development of new therapeutics, a major focus of synthetic chemistry. Education and energy have already been discussed. Environment, food, population, and water are all closely linked and are being addressed on several levels by chemists. These ten issues are those that Smalley found were the top concerns of his lecture audiences when he queried them, and they closely track the interests of college students whom this book aims to inspire. Faculty teaching chemistry, especially at the introductory levels, should find creative ways to work these issues and their potential solutions into their lectures.

The subject of chemistry and energy is specifically addressed in more than one letter to Angela; and the last three chapters, including the one by Uncle Carl, focus specifically on this topic. Interesting quantitative data are provided about the amount of power from the sunlight that reaches the earth, its relationship to the global energy needs of our planet over the next several decades, and present and future technologies and science devised by the chemistry community to capture and utilize solar energy. A "Powering the Planet" program launched by chemistry faculty at MIT, Caltech, and other institutions to address this issue is described. An especially attractive idea, one that I hear about often from my colleague Dan Nocera, involves the "splitting" of water into hydrogen ($H_2$) and oxygen ($O_2$) by sunlight with appropriately designed catalysts. This energetically uphill reaction stores solar energy in the H—H and O—O chemical bonds of these two molecules for subsequent release and utilization when they are recombined in a fuel cell to produce water, an environmentally friendly compound, and electricity. The process bears a striking resemblance to the use of sunlight in nature by green plants, which use water, $CO_2$, and solar photons to make hydrocarbons and dioxygen, which are combined in animals to supply energy with the release of $CO_2$ and water back into the environment. Chemists are working intensely to produce new catalysts for the efficient capture of sunlight, to convert solar energy to electrical energy, and to employ the latter to convert water to hydrogen and oxygen as a means of storing that energy for subsequent conversion to water and energy at a time when and/or a place where sunlight is not available. Other strategies described in the letters to Angela reinforce some of the fundamental ideas and the discussions of the life sciences and materials written in earlier chapters of this book. I now turn to these other sections.

The most important decisions in life are often made through an emotional rather than intellectual process; if the choice "feels right" it is usually taken. But the individuals whom one encounters along the way and intensely personal experiences often influence the choice of career and, for scientists, of research direction. Letters by Marye Anne Fox, Abhik Ghosh (Angela was an exchange student in his lab), and Terry Collins nicely illustrate how specific people fueled their interest in fundamental chemistry and, eventually, its applications. A fascinating letter by the Sessler brothers, Jonathan and Daniel, spins a tale in which our fictional Angela had fallen ill and required surgery when she was 7 years old. Dan, an anesthesiologist resident at UCLA at the time, saved Angela from a rare, life-threatening problem that occurred during the procedure. The letter describes how many anesthetics work to induce the unconscious state and mask pain, detailing fundamental principles of chemistry often taught at the freshman level. One can only wonder whether Angela's own personal experience, recounted in the letter by the Sesslers, might inspire a career in chemistry applied in the medical sciences. A letter by Chaitan Khosla describes how diseases diagnosed in his wife and son led to his interest to investigate an auto-immune disease brought about by eating gluten.

The interface between chemistry and the life sciences is broad and deep. Six of the letters to Angela address specific areas for study at this interface, and several in the other sections of the book contain related information. Letters by Judith Klinman and Marie-Alda Gilles-Gonzalez regale us with the role of dioxygen in biochemistry, specifically how nature has evolved systems to detect and utilize it. Two letters by bioinorganic chemists Liz Nolan (Angela's cousin) and Kara Bren describe the biological detection of metal ions involved in cell signaling, such as calcium and zinc. They also recount the use of metal ions to probe human health by MRI and gamma radiography and to treat diseases such as arthritis and cancer.

Another important and topical interface is the one between chemistry and materials science including nanotechnology. Three letters involving biomaterials, supramolecules, and nanoparticles comprise this section. An engaging letter from Michael Sailor begins with a reference to physicist Richard Feynman's lecture entitled "There's plenty of room at the bottom," which some view to have heralded the emergence of nanotechnology. Sailor then recounts the Doctor Seuss book *The Cat in the Hat Comes Back!* in which the concept of an indivisible limit is cleverly introduced for children. He then discusses GORE-TEX, the outdoor clothing fabric that repels liquid but not gaseous water as a further entrée into many nanotechnological feats including quantum dots, silicon-based photonic crystals, and "nano-worms" used to find a tumor in the body.

In closing, it seems appropriate to note that the publishing of this book in 2011 coincides with the International Year of Chemistry, a time when the world celebrates the "achievements of chemistry and its contributions to the well-being of humankind," as articulated on the official IYC 2011 web site. Whereas IYC 2011 looks backward, *Letters to a Young Chemistry* is a forward-looking collection, providing motivation and encouragement to the next generation of chemists on whom society must rely to help solve the problems and provide the breakthrough discoveries required to meet the 10 Smalley issues and many others. The community owes a debt of gratitude to Abhik Ghosh for his inspiration in conceiving this project and his persistence in assuring that the letters be assembled within a reasonable time frame, especially given the busy schedules of all the writers. Surely they, like I, share Abhik's passion for chemistry and its future. We hope that the community of readers, including students, educators, and the public at large, will find the efforts worthwhile.

# *Preface*

Walk into a major bookstore in your area and wander over to the popular science section, as I have done countless times in the last few years, mostly in the US. You'll likely find several shelves devoted to physics and to biology. By contrast, the collection on chemistry is likely to be small, almost pathetic, consisting of little more than a book or two about Marie Curie (the centenary of whose chemistry Nobel Prize falls in 2011, the International Year of Chemistry), the periodic table, and the like. Important as these topics are, they hardly convey anything about the excitement that contemporary chemists feel about their subject, and especially about its future. Nor does the popular literature do justice to the massive contributions that chemistry has made and continues to make to human welfare. This book is an early effort to fill a major gap in the popular science literature.

Misperceptions about chemistry's significance and role abound. Consumer products are often advertised as "chemical-free", as though healthy products are not made of ordinary matter. Ignorance about chemistry, however, goes beyond public concern (much of it well-justified) about pollution, carcinogens, mutagens, etc. The respected magazine *The Economist* provides a number of examples that would be rather hilarious were it not for their egregious nature. Because the periodic table is essentially complete, except for some fleeting, super-heavy elements, the magazine contends that chemistry has "lost its oomph." Thus, in recent years, whenever the chemistry Nobel Prize has gone to a topic on the chemistry-biology interface, *The Economist* has viewed the matter as the Nobel Committee's way of awarding two prizes in biology. Imagine the shock felt by the magazine's science staff when not only the 2010 Nobel Prize in chemistry went to fundamental developments in organic synthesis (palladium-catalyzed coupling reactions), but the physics Nobel also went to a chemically oriented theme, namely the discovery of graphene. Well, enough about *The Economist*; I simply wanted to illustrate one of the more serious misperceptions that are out there. Sadly, there are many more.

Unfortunately, it's not just science journalists who are sometimes inadequately informed about chemistry. Physicists and biologists often ask us: What are chemistry's grand challenges? There are many. Some years ago (*Chemical and Engineering News,* August 7, 2000), Professor Stephen Lippard assembled a list of some twenty such grand challenges (indeed this is part of the reason I requested Steve to write a Foreword for this book), while freely admitting that there were many more. Unlike the physicists, we do not seek theories of everything, but that's nothing to be ashamed of. It's a sign that our science is healthy, vibrant, and brimming with exciting problems that will challenge the brightest minds for generations to come. Even practitioners of closely related fields such as biochemistry sometimes fail to see the value of chemistry's fundamentals. Not long ago, I read an interview of a famous structural biologist in an equally famous journal, where he stated that subjects like inorganic chemistry had simply ceased to exist! Ninety-five percent of the periodic table is no longer worth studying? My first reaction was that the contention was too ridiculous to merit a rebuttal. Yet rebut we must. Several chapters in this book do just that, i.e. show that the study of the fundamentals of chemistry is alive and well.

Why is chemistry so poorly represented in the popular media? This is not an easy question to answer because causation is often difficult to pinpoint. What is clear, however, is that chemists haven't quite made the effort—on the same scale as physicists and biologists—to bring the excitement of the molecular world to the public. Chemistry's unique iconic language has been seen as an impediment, as has the subject's somewhat detail-oriented nature. Finding a transition metal reagent that activates (i.e. reacts with) a normally inert C-H or C-F bond is an utterly fascinating exercise that is no less deserving of a bright young person's interest than, say, the origin of the universe. Chemists must bite the bullet and take the trouble to better explain what they do and why it's scientifically worthwhile and personally rewarding. If the public finds it fascinating to read about black holes and string theory—hardly accessible stuff, by anyone's definition—surely a handful of them will find it of interest to read a popular article (or even a book) about C-H activation, to pick a somewhat random example of an important chemical problem.

In the 17 chapters that follow, my fellow contributors and I have tried to fulfill precisely this goal, namely, to explain serious chemistry research in accessible language. We have done so in the form of letters to a hypothetical young girl, Angela, a UCSD undergraduate who has written to us requesting information on career opportunities in different areas of chemistry. The detailed responses of her correspondents form the body of the book. Steve Lippard's broad vision and sense of where chemistry is going made him an ideal person to write the Foreword, an invitation he graciously accepted. In his Foreword, he provides a sampler of what the reader can expect to find in this book. Given the excellent job done by Steve, I will refrain from commenting on the individual chapters. Instead I'll dwell briefly on the origin of the book.

Books with the title *Letters to a Young XYZ* have become a genre of their own. The ur-*Letters* book is of course Rainer Maria Rilke's *Letters to a Young Poet*, but there are several others that are both instructive and enjoyable. Publication of Ian Stewart's *Letters to a Young Mathematician* a few years ago made me sorely miss a similar book for chemistry students and led me to envision the present volume. Around the same time appeared Natalie Angier's *The Canon: A Whirligig Tour of the Beautiful Basics of Science* and Bill Bryson's *A Short History of Nearly Everything*. I loved these two books, as much for their substance as for their literary styles. I wanted to do for chemistry something akin to what Angier and Bryson had done for all of science. Unfortunately, I was in no position to do so. With a full teaching load and a fair-sized research group, I had little time to write a book, let alone one on popular science. I did the next best thing: I turned to my friends and colleagues, including many on the US West Coast (where I came of age as a chemist and where I still visit on a regular basis), telling them of my idea of a *Letters* book and asking them whether they would consider contributing a chapter. I was humbled by the universally positive response. Elder statesmen and young assistant professors alike, as well as everyone in between, gladly took time out of their busy schedules to contribute to the book. Evidently, the urge to teach and touch the next generation of scientists in a positive way is a strong one. Unlike other *Letters* books, this book is thus a multiauthor effort. We may lack some of Angier's and Bryson's literary flair, but as practicing scientists, we bring a first-hand account of science, which should be distinctive and attractive on its own merits.

A few words about Angela, the protagonist of the book. Her character is a pastiche of a number of young students—both young men and women—I have met and taught over the years. As I picture her, she comes from a modest to average social background; her mother is a high-school English teacher, her father a lawn care professional. The eldest of three siblings, she remembers well some of her family's financial hardships during her youth and therefore appreciates all the more her current status as a UCSD undergraduate. Enthusiastic and bright, she has been an avid participant in undergraduate research. Outside of school she enjoys spending time with friends and family and the fabulous outdoors of southern California. In many ways, she is like thousands, if not millions, of her generation worldwide. She loves science but also appreciates the importance of a well-rounded life. A number of the authors, myself included, found it natural to claim personal acquaintance with her (in our slightly make-believe world). Her position is not unlike that of a child, a grandchild, a sibling, a cousin, a favorite student, etc. for many of us. When writing my own chapter, I wondered again and again whether I could give the same advice to my fifteen-year-old son, who is now in high school, finds science fascinating but, like Angela, is unsure whether and how it can be a part of his life. I can honestly say that the answer is a resounding yes. I am sure that the same can be said of all my coauthors.

So what exactly makes chemistry a career worth pursuing? Admittedly, it's not the easiest of careers, but it's an amazingly exciting and fulfilling one for many people. As mentioned in Steve's Foreword and in a number of Chapters, chemistry is key to solving many of today's most pressing problems such as disease, energy, and food supply. At the same time, it would be wrong to view chemistry as simply a kind of central service for the scientific-technical enterprise. Stunning discoveries in the most fundamental aspects of our science are taking place at an ever-accelerating pace. Have a quick browse through a current issue of *JACS* or *Accounts of Chemical Research*, or better still (if possible), attend a national meeting of the American Chemical Society and listen to some keynote lectures in different areas. You'll be left with no doubt that chemists work and live in a world of breathtaking discoveries that are transforming the way we view and deal with the molecular world—and indeed life itself.

Because of the manner in which this book developed, the majority of the authors of this book are US-based. The broad subject matter, however, should be of international interest. Whether you are reading this book from North America, Europe, Australia, or for that matter Brazil, China, India, Russia or other emerging country, you should be able to identify with the hopes and dreams implicit in this book. But what if you love science but are located in a poorer part of the world, where meaningful scientific research is not possible? Admittedly, that's tough and I am in no position to offer magic bullets. There are hopeful signs all around, however. By and large, the developing economies are

growing quickly; the Internet is deeply empowering. There are scholarship opportunities abroad. Few things are more satisfying to me than
the fact that some of my best students and collaborators have come
from Africa, including some from truly disadvantaged backgrounds.

To conclude, I wish to thank again the chapter authors for generously
contributing their time and sharing their visions, as well as for their
faith in my ability to pull through this project. The book has been carefully but lightly edited so as to preserve the authors' original viewpoints
and styles; indeed, occasionally the authors have expressed opinions
that differ considerably from my own. Steve Lippard read every chapter
before composing his thoughtful Foreword. The individual chapters
have also been checked for grammar and style by three of my collabo-
rators/students, Dr. Adam Chamberlin, Mr. Hans-Kristian Norheim and
Mr. Simon Larsen—my sincere thanks to them all. I cannot express my
appreciation strongly enough for Anita Lekhwani, a Senior Editor at
Wiley. My ideas found fertile ground the moment I mentioned them to
her and, since then, her suggestions and words of encouragement have
been a morale-booster throughout the editorial process. Wiley's Senior
Production Editor Christine Punzo and Project Manager Janet Hronek,
consummate professionals, made the production process painless and
even enjoyable!

I hope you find this book useful and enjoyable.

Abhik Ghosh

# Contributors

KARA L. BREN, Professor, Department of Chemistry, University of Rochester, Rochester, NY 14627-0216 USA; Email: bren@chem.rochester.edu

PENELOPE J. BROTHERS, Professor, Department of Chemistry, The University of Auckland, Private Bag 92019, Auckland, New Zealand; Email: p.brothers@auckland.ac.nz

CYNTHIA J. BURROWS, Professor, Department of Chemistry, University of Utah, 315 S. 1400 East, Rm. 2020, Salt Lake City, UT 84112-0850 USA; Email: burrows@chem.utah.edu

DAVID A. CASE, Professor, Department of Chemistry and Chemical Biology, and BioMaPS Institute, Rutgers University, 610 Taylor Road, Piscataway, NJ 08854-8066 USA; Email: case@biomaps.rutgers.edu

SETH M. COHEN, Professor, Department of Chemistry and Biochemistry, University of California, San Diego, 9500 Gilman Drive, La Jolla, CA 92093-0358 USA; Email: scohen@ucsd.edu

TERRENCE J. COLLINS, Professor, Department of Chemistry, Carnegie Mellon University, 4400 Fifth Avenue, Pittsburgh, PA 15213 USA; Email: tc1u@andrew.cmu.edu

MARYE ANNE FOX, Professor, Chancellor's Office, University of California, San Diego, 9500 Gilman Drive, MC 0005, La Jolla, CA 92093-0005 USA; Email: mafox@ucsd.edu

ABHIK GHOSH, Professor, Department of Chemistry, University of Tromsø, 9037 Tromsø, Norway; Email: abhik.ghosh@uit.no

MARIE-ALDA GILLES-GONZALEZ, Professor, UT Southwestern Medical Center at Dallas, 5323 Harry Hines Blvd., Dallas, TX 75390-9038 USA; Email: marie-alda.gilles-gonzalez@utsouthwestern.edu

HARRY B. GRAY, Professor, Beckman Institute, California Institute of Technology, Pasadena, CA 91125 USA; Email: hgcm@its.caltech.edu

CHAITAN KHOSLA, Professor, Department of Chemistry, Stanford University, Stanford, CA 94305-5080 USA; Email: khosla@stanford.edu

JUDITH P. KLINMAN, Professor, Department of Chemistry, University of California, Berkeley, 608 Stanley Hall, QB3 Institute, Berkeley, CA 94720-3220 USA; Email: klinman@berkeley.edu

STEPHEN J. LIPPARD, Department of Chemistry, Massachusetts Institute of Technology, Cambridge, MA 02139 USA

JOHN S. MAGYAR, Professor, Department of Chemistry, Barnard College, 3009 Broadway, New York 10027 USA; Email: jmagyar@barnard.edu

ELIZABETH M. NOLAN, Professor, Department of Chemistry, Massachusetts Institute of Technology, 77 Massachusetts Ave., Cambridge, MA 02139 USA; Email: lnolan@mit.edu

MICHAEL J. SAILOR, Professor, Department of Chemistry and Biochemistry, University of California, San Diego, 9500 Gilman Drive, La Jolla, CA 92093-0358 USA; Email: msailor@ucsd.edu

DANIEL I. SESSLER, Professor, Department of Outcomes Research, The Cleveland Clinic, Cleveland Clinic Main Campus, Mail Code P77, 9500 Euclid Avenue, Cleveland, OH 44195 USA; Email: ds@or.org

JONATHAN L. SESSLER, Professor, Department of Chemistry and Biochemistry, The University of Texas at Austin, University Station A5300, Austin, TX 78712 USA; Email: sessler@mail.utexas.edu

CARL C. WAMSER, Professor, Department of Chemistry, Portland State University, Portland, OR 97207-0751 USA; Email: wamserc@pdx.edu

JONATHAN J. WILKER, Professor, Department of Chemistry, Purdue University, 560 Oval Drive, West Lafayette, IN 47907-2084 USA; Email: wilker@purdue.edu

# Part I

# *From Fundamentals to Applications*

# 1

# Let's Get Physical

**Marye Anne Fox**

*University of California, San Diego*

Marye Anne Fox is a distinguished professor of chemistry and the seventh chancellor at the University of California, San Diego (UCSD). She received her Bachelor of Science in Chemistry at Notre Dame College and her PhD, also in Chemistry, at Dartmouth College. After a National Science Foundation (NSF) postdoctoral appointment at the University of Maryland, she joined the faculty at the University of Texas, where she was appointed ultimately to the Waggoner Regents Chair in Chemistry and was named Vice President for Research. She served as Chancellor and Distinguished University Professor at North Carolina State University before assuming her present position. She has been recognized nationally and internationally for her contributions to chemical research, science education, and innovative service to higher education. She is a member of the U.S. National Academy of Sciences, the American Academy of Arts and Sciences, and the American Philosophical Society, and has served on the National Science Board and the President's Council of Advisors on Science and Technology. She has received 12 honorary degrees.

*Letters to a Young Chemist*, First Edition. Edited by Abhik Ghosh.
© 2011 John Wiley & Sons, Inc. Published 2011 by John Wiley & Sons, Inc.

Dear Angela,

I am delighted to hear that you had a great summer working in a chemistry lab! And combining that experience with study abroad in Norway, which is such a beautiful place, must have been truly wonderful! Quite a change for a Southern California girl.

I assume you had a very positive experience, and I suspect it also means you've been bitten by the research bug. By that, I mean that you've probably had that indescribable feeling that comes when you've synthesized a molecule that never existed before or when you were the first in human history to understand why a particular reaction takes place the way it does. Nothing like it, really. And once you've experienced it, you'll find it hard to live a life that doesn't include the possibility of discovery. The pursuit of new ways of thinking about nature is addictive, and getting to the goal by proving or disproving your original proposal about what might happen will keep you working long hours for months or years at a time. Your friends may think you're crazy to work all night on occasion, but you know something they'll never understand.

I'm also glad to hear that you've chosen to pursue your degree. That shows a lot of good sense, given the proximity of your family at UCSD and your determination to make scientific research a big part of your life. Quite aside from the natural beauty of the campus and the fact that San Diego has the world's best weather all year long, UCSD is a fabulous place to undertake serious scientific studies. *Newsweek* magazine announced a couple of years ago that UCSD is the "hottest place to do science" in the United States. One indicator of the quality of our scientific research programs is that the 2008 Nobel Prize in Chemistry was awarded to one of our long-time faculty members.

As is true in most research universities worldwide, you'll have a chance to get involved in research right from the beginning. Once you demonstrate that you want to do serious research, you'll likely be able to work with faculty who are well funded by federal granting agencies, and some will be able to offer stipends to students who work half-time or so with the research group on their research projects. As a result, working on a project you love can also help you address the costs of attending a premiere college or university.

One of our highest priorities at UCSD has always been to involve undergraduates in research as soon as they can demonstrate that they can contribute to a particular research group's efforts. Having worked in a lab for a summer already, you should be ready as soon as you arrive on campus to connect with one or more professors whose research work interests you.

## MY LIFE AS A RESEARCH CHEMIST

I think it's a great idea that you want to find out a bit more about opportunities in different areas of chemistry so you can make a more informed choice about the direction you'll ultimately pursue. My area of choice has been physical organic chemistry, with specialization in photochemistry, electrochemistry, and materials chemistry. Virtually all of our work is aimed at understanding how a particular reaction takes place, often in great detail. It includes the scientific question of how changes in structure induce changes in chemical reactivity, in the ground state or the excited state of a molecule or a family of molecules.

Given that reactions take place through a series of bond-making and bond-breaking steps, this work often involves reactive intermediates. A full description of a reaction also defines the rates of reaction (kinetics) and the energy changes encountered as the reaction proceeds (thermodynamics). This is exactly where organic chemistry interfaces with physical chemistry. This description of a chemical reaction, in fact, defines the field of physical organic chemistry. A key step in accomplishing such a description consists of characterizing the electron flow and the reactive intermediates forming along a reaction pathway. In a larger sense, physical organic chemistry defines how a local environment can affect reaction rates by influencing the stability of a key transition state. In many cases, it is also possible to make models of transition states with the help of theoretical calculations and to use theory to predict evolving chemical reactions. It's extremely rewarding intellectually to work with theorists to establish by inference how a series of chemical bonds are broken and formed, and hence to be able to devise new chemical transformations.

Although I spend a great deal of my time as a chancellor at UCSD, I am still a chemistry professor at heart. In that capacity, I've worked with a highly talented group of students to study a variety of physical organic problems. I'm especially proud that we were able to make a major contribution toward defining a new field of organic photoelectrochemistry, which involves a combination of surface chemistry and excited state chemistry. This work involved syntheses of new molecules, in ground and excited states, and observing how their structure affects subsequent chemical reactivity. We also carried out theoretical calculations in order to test our interpretation of experimental observations. We are often interested in being able to predict physical properties for compounds or materials that don't even exist until we make them.

I'm happy to say that over 60 students have completed advanced degrees in my research group, and many of them have themselves established well-regarded research programs of their own. Because of their hard work, I was elected at a relatively young age to membership in the National Academy of Sciences, which is a profound honor for any scientist. I really miss the days when I could spend most of my time with students in the lab.

## LEARNING TO DO RESEARCH: PHYSICAL ORGANIC PRINCIPLES

Perhaps it would be useful to you to learn how I came to develop an interest in the physical properties of organic molecules. It was basically a series of very positive research experiences that led me to become a research chemist, and thereafter an independent faculty member at a research-focused university.

My first exposure to chemistry research was as a student in a summer program supported by the NSF at the Illinois Institute of Technology in Chicago. As a student in the program called Research Experience for Undergraduates (REU), I was assigned to Professor Jerry Kresge's research group, thereby working side by side with postdoctoral fellows, visiting faculty members, and graduate and undergraduate students. I was very pleased to be accepted so cheerfully into the group, given my status as an inexperienced REU student. Professor Kresge was interested in determining how acids could effect changes in the hydrolysis of vinyl ethers. We wished to determine whether Bronsted or Lewis acidity was involved, as well as to identify key reaction intermediates encountered as the reaction took place. The work would provide important information about how acids can catalyze (i.e., accelerate) certain reactions.

My contribution to the project was to synthesize cyclohexenylethyl ether and to monitor the kinetics of its hydrolysis to cyclohexanone. The rate of hydrolysis could be followed by monitoring the appearance of the ultraviolet absorption of the ketone product. We then established reaction rates for other vinyl ethers, for example, those with smaller rings or appended functional groups, and with enhanced or reduced sensitivity toward various acids. The question was whether a general acid ($H^+$) or a specific acid (HX, where $X^-$ is a counterion that is involved in the key transition state) induces the catalytic acceleration of the observed rates. In turn, this allowed us to find out exactly how this reaction proceeds. Ultimately, the work led to my first paper in the *Journal of the American Chemical Society*.

Having had a very positive experience in a physical organic group, I loved the idea of undertaking simple syntheses to make new compounds in which fundamental changes in reactivity could be brought about by changes in structure. So, when I had the chance to work with Professor Roger Binkley on problems involving photochemical excitation, I jumped at it. Photochemical reactions are those that take place after the absorption of light. Photochemical excitation is perhaps the easiest way of inducing a reactivity change with minimal structural change. Photochemistry is therefore an exceptionally important subarea of physical organic chemistry.

Although Dr. Binkley was interested in carbohydrate photochemistry, I chose to work on a structurally simpler compound, benzalazine. We wanted to measure the relative rates of cleavage of the C–N bond by monitoring the quantum yields for consumption of the starting material and for the appearance of product. The underlying goal of this work was to determine the multiplicity of the excited state leading to each product; that is, we wanted to know whether a singlet or triplet state was involved. The insight afforded by establishing the reaction kinetics has profoundly influenced my research for the rest of my career.

Perhaps even more important, the Binkley research group met each Wednesday night at his home to have a simple dinner in which the dessert consisted of working together to solve a tantalizing mechanistic problem. Sometimes, the problems involved reactive intermediates, often radicals or diradicals produced by photochemical excitation. As a beginner, it was hard to imagine how a molecule could be twisted so much to yield highly strained or rearranged compounds. In general, we practiced arrow pushing to determine the electron flow that defines a reaction mechanism.

Moving to Hanover, New Hampshire, to attend graduate school at Dartmouth College was one of the most important decisions of my life. Not only did I find a brilliant mentor, Professor David Lemal, a person who cared equally about teaching excellence in research and about his students, but because the program was small, I was also able to work frequently with other faculty working on quite different projects. I recall fondly working with Professors Walter Stockmeyer (a polymer chemist), Chuck Braun (a physical spectroscopist), Tom Spencer (a biochemical kineticist), and Gordon Gribble (a synthetic organic chemist). Because of their invaluable insights, I was able to finish my doctoral work quickly, in 3 years, so I could join my husband in Washington where he was assigned after having been drafted into the U.S. Air Force.

The project I began in graduate school was to prepare perfluoro-tetrahedrane, a highly strained compound in which four carbon atoms were arranged so as to resemble a pyramid with a fluorine atom at each corner. The molecule was bound to be highly strained and to have interesting physical properties. I worked for nearly a year pre-paring various precursors and applying what I had already learned about photochemistry. I discovered some interesting routes to strained compounds, but the goal remained elusive. This was the first time I had to deal with failure in the lab. But even in this failure I learned an important lesson: when a project is worthwhile, there will be chal-lenges, and sometimes it will be useful to fail quickly and get on to another project. I did have the chance several years later to collabo-rate with a German friend who had synthesized tetrakis(*t*-butyl)tet-rahedrane, so the project continued to be close to my heart.

In my case, my failing to synthesize my target meant shifting to studying the valence isomerization of several families of halogenated arenes. I discovered a new reactive biradical derived from chloroben-zene and several interesting interconversions among halogenated pyrazine, pyridazine, and pyrimidines, along with fully defined mecha-nisms by which the conversions took place.

Besides learning about broad areas of science, I used my graduate school experience to learn time management and how to balance various competing demands from my personal and professional life. The norms of graduate life in chemistry also reinforced a strong core work ethic that I've had throughout my life. To this day, administra-tion colleagues marvel at my work capacity, which is simply normal by chemists' standards.

I will always be indebted to my research advisor David Lemal for encouraging me to realize that children should not be forgone as part of the life of an academic, whether male or female. He was gracious in letting me work odd hours and still keep my stipend when my first son was born, in order to stay on track to achieve my 3-year comple-tion goal. In doing so, he taught me invaluable life lessons. (These lessons, even now, I have observed, are not consistently understood or practiced by many of my professional colleagues. I've found it remarkable to observe how much more efficient some of my male and female colleagues became once they took on the responsibility of parenting a child.) I hope that as you pursue your interest in science, you will be fortunate to have a research supervisor and/or colleague who will be similarly supportive at crucial junctions in your life. Special advice: choose your mentor (and your spouse!) carefully.

Next I accepted a position in a new NSF postdoctoral program Research Applied to National Needs (RANN) at the University of Maryland. Consistent with the applied nature of the program, I developed a new instrument for using fluorescence spectroscopy to monitor air quality in real time. That work resulted in my first publication in *Science*, a highly prestigious broad-interest journal. The experience also encouraged me to think deeply about directions I wanted to pursue in my independent career.

## BECOMING A CHEMISTRY PROFESSOR: CONDUCTING INDEPENDENT RESEARCH

The decision to become a professor and to pursue an academic research career is an important one. It requires a commitment to work with students to investigate areas that are (1) original, (2) interesting, and (3) important. Only if all three criteria are met will you attract federal or foundation support for your research. I was fortunate that my research group was well funded from its inception. NSF, Department of Energy (DOE), and the Welch Foundation provided reliable and continuous funding for my projects as soon as I took my first job at the University of Texas at Austin.

From the background I described above, you could probably guess that I would try to incorporate photochemistry or physical organic chemistry as major directions for my independent research program. I was interested in chemical bonding and quickly made arrangements to teach a course on molecular orbital and perturbation theory with the brilliant theorist Michael J. S. Dewar. I joined a literature seminar on Wednesday nights with four other faculty and their graduate students to work on reactive intermediates, isotope effects, molecular strain, aromaticity, pericyclic reactions, and electronic effects. We also worked on noncovalent molecular recognition and supramolecular chemistry, which was then a newly emerging field. We worked on kinetics and thermodynamics, especially on transition state theory, focusing on Hammett plots and the Hammond postulate. I contributed what I knew of acid–base chemistry and photochemical mechanisms, with a strong emphasis on catalysis, both organic and enzymatic. We taught ourselves about organic magnets, organometallic compounds and of materials such as polymers and liquid crystals. Above all, we worried about electron flow in reaction mechanisms. The experience was invaluable, and I strongly urge you, as your time permits, to find a similar group wherever you pursue your graduate degree.

I began my own work as an assistant professor by scouring the literature for important but unanswered scientific questions. I was particularly interested in those in which there had been recent progress but with less than complete success. With this goal in mind, I was fascinated to find a report by Fujishima and Honda of The University of Tokyo that illuminating an electrochemically poised semiconductor surface could split water into oxygen and hydrogen, thus making a combustible fuel from an abundant and safe precursor. The reality, however, was that the applied potential was too high and the quantum efficiency too low for practical applications. It was an ideal challenge for an assistant professor.

We hoped we could contribute to understanding the fundamentals of this important reaction while addressing its practical consequences in parallel. Accordingly, we set out to characterize photoinduced electron transfer (PET), with applications ranging from fundamental theory through very practical applications.

## PET

We have focused on environmental control of chemical reactions using the absorption of light as a trigger to allow spectroscopic monitoring of a series of bond-making and bond-breaking steps. We have been particularly interested in the excited states of species called reactive intermediates, especially those involving PET. We frequently use laser spectroscopy to follow the reaction progress on the nanosecond or picosecond time frame. When conducted in an appropriately configured multicomponent system, this work also provides practical new vehicles for solar energy utilization and storage.

Figure 1.1 presents a simple schematic of how an electron donor *D* can interact with an electron acceptor *A* and the experimental

One-electron transfer

$$D + A \rightleftharpoons D^{\cdot +} + A^{\cdot -}$$

Distance

Orientation

Solvation

Thermicity

**Figure 1.1.** *Experimental variables affecting donor–acceptor interactions.*

variables that can affect the course of this simple reaction. First, notice that radical ions are produced from neutral precursors. To understand the reaction, we must know about reactive intermediates and about how the distance between reactants and products can influence the efficiency of PET. By changing the energetics and kinetics for the forward and back electron transfers, many practical applications were uncovered. In addition, the rates of these laser-induced chemical reactions were used to define the influence afforded by the polarity of the solvent or the structure of the solid to which the *A/D* system was adsorbed. It has also led us to think in more detail about the orientation of molecules bound to surfaces.

We then began to construct more complex arrays that have two acceptors for each donor, as shown in Figure 1.2. Upon photoexcitation of either the donor or acceptor, PET takes place. The electron can then hop to the second acceptor, creating an energetic and physical barrier to back electron transfer that would dissipate the initial photonic energy. By inhibiting back electron transfer, three component arrays like this lengthen the lifetime of an oxidized/reduced ion pair. As a result, the chance of observing net photochemical change is much higher. Figure 1.2 uses a straight line to indicate the bonding between the donor and the two acceptors. This is simply intended to indicate that any linking agent could fulfill the role, so long as the $D–A_1–A_2$ sequence is attained.

What kind of intervening materials could be used? We can look to nature for models: for example, photosynthesis itself takes place by light absorption by a special pair of porphyrins rigidly held within a

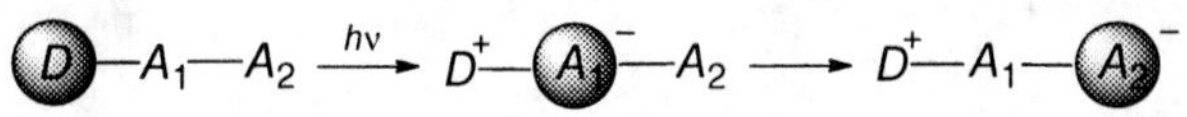

$$\textcircled{D}-A_1-A_2 \xrightarrow{h\nu} D^+-\textcircled{A}^--A_2 \longrightarrow D^+-A_1-\textcircled{A_2}^-$$

**Figure 1.2.** *Schematic representation of an electron donor–acceptor array.*

protein matrix. The initial PET then causes an electron to hop (as in Figure 1.2) to another position and then to another. This accomplishes directional electron transfer across a lipid bilayer membrane, with the electron ultimately being trapped as a reduced quinone, and finally by an iron–sulfur complex. It is impossible to understand photosynthesis, possibly the most important process for the sustainability of our planet, without a fundamental understanding of the variables that control the efficiency and directionality of PET.

Suppose that the intervening material (between the donor and acceptor) is more conformationally flexible than what is encountered in the photosynthetic reaction center, as shown in Figure 1.3. Under these circumstances, the distance between the donor and acceptor would be controlled by the movement of the intervening chain. Three situations are illustrated here: (1) a rigid backbone in which separation distance is controlled by the original synthesis, (2) a flexible backbone in which distance changes with time after pulse photoactivation, and (3) a situation in which the polarity of the backbone differs sufficiently from that surrounding the highly solvated radical ion pair that phase separation takes place. These are the same structural criteria that represent major design goals in functional polymer chemistry. As a result, the study of PET, when induced by a laser pulse, provides an excellent probe for an entirely different area of chemistry, that is, polymers and polymer blends.

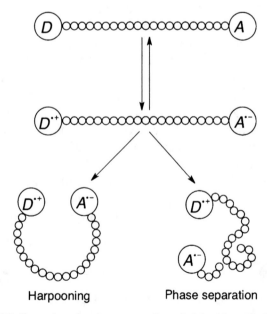

**Figure 1.3.** *Dynamics of a donor–acceptor pair linked by a flexible chain.*

   Furthermore, there is no requirement that the individual components (represented by balls in Figure 1.3) are covalently bound. In
fact, we have shown that discotic liquid crystals can similarly act as
intervening supports for PET. Liquid crystals display a long-range
order induced by noncovalent stacking of individual units. In some
cases, hundreds of component molecules are aligned, producing an
order over hundreds of nanometers. One might therefore expect differences in PET activity on both sides of the transition temperature
between one liquid crystalline phase and another.

   Peptides represent yet another family of intervening materials that
provide a probe for conformational effects. Because peptides can
aggregate as helices or as sheets, they produce a macromolecular
order comparable to that of liquid crystals. Dipole effects in helical
peptides are additive, so that a macroscopic dipole from the C- to the
N-end of the peptide (or vice versa) can enhance or oppose the
pulsed PET.

   Nor do the donor and acceptor need to be equally soluble in polar
or nonpolar solvents. Consider the situation if the donor were a
nanoparticle surrounded by an acceptor strongly adsorbed on its
surface, as depicted in Figure 1.4. In such a shell–core composite,
light absorption would initiate charge separation in a pathway directly

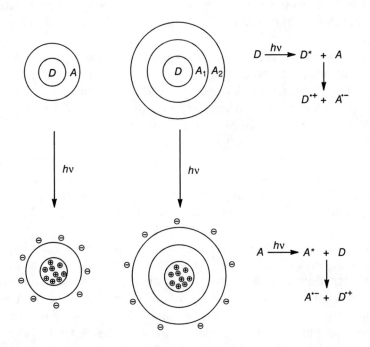

**Figure 1.4.** Photodynamics of acceptors adsorbed on a donor nanoparticle.

comparable to that seen in the linear arrays. Spectroscopic probes for lifetimes of the charge-separated pairs and for direct measurements of the rate of back electron transfer will be critical measurements before these size-differentiated materials can be included as manufactured components of nanodevices. As nanoscience progresses, it is likely that pulsed laser-induced PET will become a major characterization technique for defining physical properties as a function of size and surface composition.

## PHOTOELECTROCHEMISTRY

The above considerations also apply to situations in which a donor and an acceptor are both reversibly adsorbed on a common surface. For example, it has been known for over a century that molecules adsorb and desorb reversibly on electrodes, producing a complex nonhomogeneous layer near the electrode surface with a substantially different composition relative to the bulk solution with which it is in contact. Electrochemists refer to this space as an electrode surface region, or a double layer, if the contacting solution is an aqueous electrolyte. Chemical reactions take place with ease in this region as the electrode potential is scanned through the oxidation or reduction potential of the reactant. After the redox reaction has taken place, equilibration takes place with the neighboring bulk solution.

The electrochemical cell in which these transformations take place consists of an anode where oxidation reactions take place, a cathode where the counterreduction occurs, and a reference electrode that allows for measurement of applied potential. Within the last several decades, it has been discovered that a potential difference can be attained not only by using a battery or a potentiostat but also by using light to produce an electron–hole pair by the photoexcitation of a doped semiconductor. This kind of cell is called a photoelectrochemical cell, and its catalytic behavior is determined by the oxidation potential of the hole and the reduction potential of the electron.

Substantial work has been done to study the properties of the doped semiconductors employed, as there was general consensus for several decades that such cells might be important for solar energy conversion and storage. Much of our work in photoelectrochemistry was done with my colleague Allen Bard. Together we wrote an article arguing that photoelectrochemical cells could be viewed as the "Holy

Grail" for electrochemists, potentially leading to efficient water splitting to produce fuel for the hydrogen economy.

Every semiconductor, by definition, has a filled valence band, separated by an energy difference (band gap) from the vacant conduction band. In an undoped state, conductivity is observed only by thermal population of the conduction band by electrons sufficiently energetic to jump across the band gap to the conduction band. This is an energetically unfavorable process and occurs only infrequently. An undoped semiconductor therefore has very low conductivity and is generally considered to be insulating.

When doped with electron-rich species, however, electrons are placed into the conduction band, which has the effect of reducing the potential of the band edges, so that the excess electrons move away from the surface of the semiconductor into the bulk. The material thus becomes conductive, and an electrode made out of such a material can function as a normal electrochemical component.

We became fascinated by potential modifications of such cells. In your mind, try a thought experiment: envision a photoelectrochemical cell in which the wire connecting the semiconductor anode (perhaps a metal oxide like $TiO_2$) and the metallic cathode (perhaps platinum, which is a good hydrogen evolution catalyst) is shrunk until the counter electrodes are in direct contact. Now fragment the electrode so it is a powder, which is depicted in Figure 1.5.

Now provide photons rather than dopants to the semiconductor surface. When the $TiO_2$ particle absorbs light of an energy greater than the band gap (wavelengths shorter than about 400 nm), an electron is promoted from the valence band to the conduction band. This absorption thus produces an electron–hole pair. This pair has an

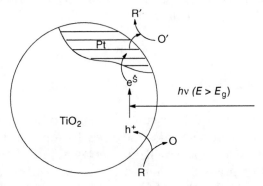

**Figure 1.5.** A platinized titania particle as a photoelectrochemical cell. O symbolizes the oxidized form of a generic organic compound, R.

oxidizing power equivalent to the band gap of the semiconductor and can strip an electron from any compound with an oxidation potential less positive than about +2.0 V (relative to the standard hydrogen electrode). This category includes essentially any organic compound containing either a heteroatom or any form of conjugation.

The photoexcitation also places an electron into the conduction band, whose redox potential is virtually identical to that of the reduction potential of adsorbed oxygen. The conduction band electron is thus converted to adsorbed superoxide. Trapping an electron–hole pair thus starts a reaction cascade between the adsorbed organic cation radical and superoxide and its cleavage products. This route, which we called organic photoelectrochemistry, allowed us to discover a wide array of new reactions that emanate from the inherent reactivity of an adsorbed cation radical. An early example of how this takes place can be seen in a 1981 *Journal of the American Chemical Society* article (see Further Reading), our first on organic photocatalysis, where we reported a fairly high-yield aerial oxidation of olefins to carbonyl compounds.

By employing different semiconductor powders suspended in a series of solvents with and without deposited metal islands, we soon developed a whole new subarea of mechanistic/physical organic chemistry. Others also joined in and expanded the repertoire of photocatalytic conversions. In addition to solar splitting of water, as noted above, this general approach has been used for detoxification of pollutants, for cleaning oil spills, as a hospital-based antibacterial, for water purification, for cleaning office windows, and much more. Suspended semiconductor powders have been used in conjunction with dye sensitizers bound or adsorbed to the semiconductor surface, as single molecules, as aggregates, or even as dye-loaded dendrimers. It's been extremely rewarding to see so many useful chemical transformations follow from such simple principles.

## SOME FINAL ADVICE

Angela, I assume you'll be writing to some other chemists as well and I hope they will also take the time to tell you in some detail about their work. I hope you'll take away some of the enthusiasm that is typical of practicing chemists. Yes, chemists work long hours, but that's generally because they love what they do. Because chemistry is a central science, it is easy to apply what you have learned and will learn in a variety of settings. With a chemistry background, you can easily move to the life sciences, to medicine, to law, to public policy, to management, to teaching, to public service, and to so many others. Who knows? You may even end up as a chancellor!

I just wish my schedule still permitted me to take on undergraduate and graduate students instead of only postdoctoral fellows. I'd surely try to make space for someone as committed as you seem to be. Best wishes for a successful stay at UCSD. I'll look forward to giving you your degree a couple of years from now.

Sincerely,

Marye Anne

## FURTHER READING

Anslyn, E. V.; Dougherty, D. A. *Modern Physical Organic Chemistry*, University Science Books, San Francisco, CA, 2006.

Bard, A. J.; Fox, M. A. Artificial photosynthesis: Solar splitting of water to hydrogen and oxygen. *Accounts of Chemical Research* 1995, *28*, 141–145.

Fox, M. A. Fundamentals in the design of molecular electronic devices: Long range charge carrier transport and electronic coupling. *Accounts of Chemical Research* 1999, *32*, 201–207.

Fox, M. A.; Chen, C. C. Mechanistic features of the semiconductor catalyzed olefin-to-carbonyl oxidative cleavage. *Journal of the American Chemical Society* 1981, *17*, 6757–6759.

Fox, M. A.; Dulay, M. Heterogeneous photocatalysis. *Chemical Reviews* 1993, *93*, 341–357.

# 2

# In Silico: *An Alternate Approach to Chemistry and Biology*

**David A. Case**
*Rutgers University*

David Case grew up in Ohio and did undergraduate studies at Michigan State University. He received a PhD in chemical physics from Harvard and has held faculty positions at the University of California, Davis, The Scripps Research Institute, and Rutgers University. His work centers on molecular dynamics (MD) simulations of proteins, nucleic acids, and carbohydrates, and he is the leader of the development team for the Amber suite of computer codes (see http://ambermd.org/). Current research interests include interpretation of nuclear magnetic resonance (NMR) results on biomolecules, the structures and mechanisms of metalloenzymes, and the development of implicit solvent models for biochemical simulations. More details are available at http://casegroup.rutgers.edu.

*Letters to a Young Chemist*, First Edition. Edited by Abhik Ghosh.
© 2011 John Wiley & Sons, Inc. Published 2011 by John Wiley & Sons, Inc.

Dear Angela,

It's a pity we didn't get a chance to meet in person. I used to live in San Diego and moved to New Jersey only last fall. Before I begin in earnest, however, let me say that I'm really impressed by your curiosity about science in general and I'll try to address your question as to whether theory and computing is something you should be taking an interest in. The world of theoretical and computational chemistry can be a lot different from the "wet labs" that most people associate with chemistry. I work in an officelike environment, with a computer screen, a messy desk, and a bunch of books as companions. Like a number of my colleagues, I ended up as a "theoretician" partly because I really enjoyed working with computers and partly because I wasn't all that good at real lab work: I still have strong memories of starting a fire in organic chemistry lab that went all the way from lab bench to ceiling. (I was lucky to have a lab partner with a cool head and ready access to a fire extinguisher.)

But it's not just a matter of fleeing to the relative safety of an office. Theoretical chemistry, at its best, provides a unifying framework for thinking about molecules and their interactions and a mechanism for the rigorous evaluation of experiments. Being able to create a model of chemical events at a microscopic (atomic) level is a key test of real "understanding" and can be a very satisfying and practical intellectual pursuit.

## A LITTLE HISTORY

At this point, I'd better define some terms. Theoretical and computational chemistry is really a branch of physical chemistry, and its appeal can be hard to appreciate before you've had a chance to study the subject during your junior year. In my own undergraduate years (at Michigan State—too long ago!), I remember being mystified by thermodynamics, then enchanted and inspired by quantum mechanics and spectroscopy. You may find the reverse to be true, or you may not be much taken with any of it. But keep an open mind: as with most parts of chemistry, you only really understand things the second time around (or when you try to teach it!).

When I was an undergrad (yes, we had computers, but slide rules were also still being made), the term "theoretical chemistry" meant "electronic structures of atoms and molecules." You already know a fair amount about the atomic part of this: even high school chemistry students learn about s, p, and d orbitals, and how filling shells of electrons helps to explain how the periodic table works. Quantum mechanics (as it was developed from about 1915 to 1935) provides a real mathematical framework that can explain and predict both these orbital patterns in atoms and the ways in which chemical bonds are formed to make molecules. As you'll see in your classes next year, the properties of some simple systems like the hydrogen atom or the harmonic oscillator (a simple model of a diatomic molecule) can be computed exactly; for more complicated (and interesting) systems, we know the equations that should be solved, but only how to get approximate solutions. Yet the existence of such equations is an enormously liberating idea. In 1929, Paul A. M. Dirac, one of the inventors of the new mechanics, claimed that "the underlying physical laws necessary for the mathematical theory of … the whole of chemistry are thus completely known, and the difficulty is only that the exact application of these laws leads to equations much too complicated to be soluble." The application of quantum mechanics to the electronic structure of molecules is now called "quantum chemistry," and some truly powerful approximation methods have been developed for these complicated equations. All of the useful methods involve large amounts of computation, and scientists who work in this area necessarily learn a lot about numerical analysis and efficient, "high-performance" scientific computing. At their most abstruse, such calculations are really a branch of applied mathematics, but practical applications of real use to chemists are also common routine, at least for the upper part of the periodic table.

Quantum chemistry is great in helping us understand how molecules work, but in practice (as opposed to Dirac's idealization mentioned above), it deals with only one or a few molecules at a time and is most directly applicable to gas-phase situations, where molecules collide with each other only rarely. Most of chemistry and all of biochemistry take place in condensed phases (solids or liquids), where molecules are in close proximity to their neighbors and are constantly being jostled about by encounters and collisions driven by heat. We can use quantum mechanics to help understand what happens when two or three or four molecules come near each other, but some additional concepts are needed to really think about solids or liquids: even a tiny drop of water contains billions of water molecules! Here we enter the areas of "simulations" and "statistical mechanics," which I'll try to explain next.

When we think microscopically about fluids like water, the concept of temperature becomes very important. It is the thermal energy at ordinary temperatures that keeps water molecules moving around and prevents them from collapsing into a regular structure like ice. In order to describe this in a computer simulation, we include thousands of water molecules around a protein in our simulation, following their thermal motions and interactions with the protein, in our best attempt to mimic what "really" happens. A big piece of the puzzle, and a challenge for these sorts of simulations, is to properly average over all these microscopic configurations to obtain energetic quantities that can be compared to thermodynamic measurements. Statistical mechanics, a branch of physical chemistry, provides a detailed procedure for how to do this. But it's a challenge to carry out all of the averaging that is required, and large-scale computations are very common in our field. You may have heard of folding@home, a project that recruits tens of thousands of otherwise idle computers from around the world to cooperate in simulations of protein folding and other biochemical simulations. Check out their Web site, http://folding.stanford.edu, to learn more about this.

## COMPUTER SIMULATION OF LIQUIDS

Many important events in biochemistry involve the interaction of proteins with other molecules since proteins serve as enzymes (catalysts) for most metabolic processes. For many of these interactions, we know enough about the structures and forces involved at the atomic level to carry out realistic computer simulations, basically by generating numerical solutions to the classical equations of motion under the influence of forces that mimic the detailed molecular interactions. Recent advances in our understanding of the nature of chemical reactions in condensed phases, along with an improved ability to carry out realistic computer simulations of macromolecules, have led to new theoretical insights into a variety of biochemical events. These have been coupled to corresponding improvements in experimental techniques for detecting reaction intermediates and in the use of new biotechnology, such as site-directed mutagenesis, for manipulating proteins in powerful new ways.

To give you a feeling for what is going on, I will tell you a bit about some of the principles involved in computer simulations of proteins, drawing on an example from the oxygen storage protein myoglobin. The example concerns the rate at which a small ligand like oxygen can diffuse through the protein and form a chemical bond to an iron atom in the heme prosthetic group. These biochemical simulations are an extension of methods that have been used for simple liquids since automatic computation became generally available, and several introductory books give a good description of techniques and results in this field.

In an MD simulation, the classical (Newton's) equations of motion for the system of interest (e.g., a biopolymer in solution), $F_i = -(\partial U / \partial x_i) = m_i (d^2 x_i / dt^2)$, are integrated numerically. From the solution of these equations, the atomic positions and velocities as a function of time are obtained (here, $m_i$ and $x_i$ represent the mass and position of particle $i$, and $U$ is the potential energy surface, which depends on the positions of the particles in the system). The time history or trajectory of the atoms permits the computation of many interesting structural and energetic features. The key steps in the numerical solution of the classical equations of motion may be divided into two parts: the evaluation of energies and forces, and the propagation of atomic positions and velocities. Empirical potential energy functions for biological molecules include energy terms to represent chemical bonds, angles, and torsions as well as rotations about bonds and "nonbonded" interactions between atoms further apart in chemical structure. Because this mathematical model resembles a physical one in which balls (representing atoms) are connected by springs, it is often called a "molecular mechanics" potential. The general form of this potential or energy function is as follows:

$$U = \sum_{\text{bonds}} K_b (b - b_{\text{eq}})^2 + \sum_{\text{angles}} K_\theta (\theta - \theta_{\text{eq}})^2 + \sum_{\text{impropers}} K_w w^2$$

$$+ \sum_{\text{torsions}} K_\phi [1 + \cos(n\phi - \delta)] + \sum_{\text{nonbonded pairs}} \left\{ 4\varepsilon \left[ \left( \frac{\sigma}{r} \right)^{12} - \left( \frac{\sigma}{r} \right)^6 \right] + \frac{q_i q_j}{r} \right\}.$$

$$(2.1)$$

The underlying energetics of molecules is determined by the motions of nuclei and electrons, which must be described by quantum mechanics; the terms in Equation 2.1 are empirical fits based on simple functions that approximate the averaged effects of the

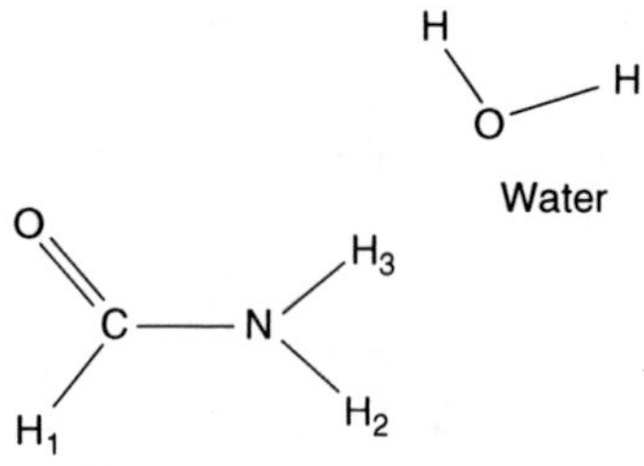

**Figure 2.1.** *Interaction of formamide with water.*

electronic motion. Figure 2.1, which shows the molecule formamide interacting with water, illustrates the nature of some of these interactions. Atoms adjacent to each other are viewed as being connected by springs (the first term of Equation 2.1), with an equilibrium distance, $b_{eq}$. Atoms bonded to a common atom (such as $H_1$–C–N) have their bond angles maintained by similar "springs" (second term of Equation 2.1). The so-called improper terms provide a restraining force to keep quartets of atoms (like $H_1$, C, O, N or C, N, $H_2$, $H_3$) in a single plane; at equilibrium, all six atoms of the formamide molecule lie in the plane of the figure. The torsional terms in Equation 2.1 describe a penalty resisting rotation about bonds such as the central C–N bond. Finally, the nonbonded terms affect all atom pairs that connect atoms in different molecules, or atoms that are separated by more than two bonds in the same molecule. The $\sigma$ values provide a size parameter for atoms, ensuring that two atoms cannot be too close to each other, while the partial charges $q$ interact with each other through Coulomb's law (the final term of Equation 2.1). Partial charges are a simple model of electrostatic effects in chemistry; for example, the molecular dipole of water is represented here as a partial negative charge associated with the oxygen atom and as partial positive charges for each hydrogen. In the configuration shown in Figure 2.1, there would be an attractive interaction between formamide and water, arising from the favorable electrostatic interaction between the partial negative charge on the water oxygen and a partial positive charge on $H_3$. When many water molecules are present, a simple description like this actually provides quite a good model for the properties of formamide in a dilute aqueous solution.

For calculations on biomolecules in water, the simulation system will typically consist of a single solute molecule (e.g., the protein) and several thousand solvent molecules in its vicinity. It turns out generally not to be sufficient to simply place a few shells of water molecules around the solute since the surface tension of the resulting water–vacuum interface can appreciably affect the properties of the system. Rather, these boundary conditions can be modified to mimic the effects of the remaining solvent that is not explicitly included, or "periodic" boundary conditions can be employed in which the system is replicated in three dimensions so that no water–vacuum interfaces remain. Because of the long-range nature of electrostatic interactions, it can be difficult to establish innocuous boundary conditions, and the question of how best to carry out solvated simulations is still an active area of research.

The internal energy $U$ is thus a complicated function of the configuration of the many-atom system, whose gradient, with respect to the Cartesian positions of each atom, provides the forces for numerical solutions of Newton's equations. The first MD simulations of proteins were carried out in the late 1970s, and the early days saw an emphasis on very fast kinetic measurements (on the picosecond or nanosecond timescale) that could be directly compared to the simulations. It was quickly realized, however, that in addition to this time-dependent behavior, the statistical properties of the configurations visited during even a short simulation can be related to time-independent thermodynamic quantities, as I discuss next.

## ESTIMATES OF FREE ENERGY PROFILES

MD simulations involve numerical determinations of individual trajectories, that is, solutions to Newton's equation of motion for particular initial conditions. In chemistry, this technique began with the study of individual gas-phase collisions, where large numbers of trajectories were run to explicitly average over initial conditions. For macromolecules or liquid simulations, however, the frequency of atomic collisions becomes so great that simulations often appear to be ergodic, such that a single trajectory samples phase space with the same distribution as do multiple simulations with randomized starting points. This implies that a dynamics simulation can be used to explore phase space and make connections to classical thermodynamics and kinetics.

I understand that you have not yet formally studied physical chemistry, so that some of these concepts may seem pretty foreign to you. But I would like to outline one sort of common calculation we do,

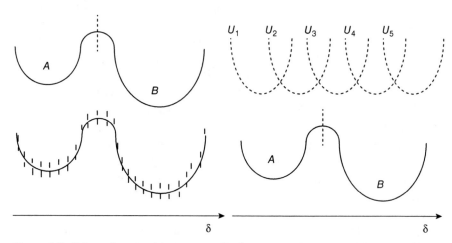

**Figure 2.2.** *Schematic potential energy profile for a dynamic system with two conformers (top left). Generalization to divide space into small bins to get* $\rho(\delta)$ *(bottom left). A series of parabolic umbrella potentials (right); $U_1$–$U_5$ could be used (in separate calculations) to force sampling of all relevant regions of the reaction coordinate* $\delta$.

including some of the math, to give you a flavor of what "chemical theory" at an atomic level looks like. If you find it fascinating to think about how equilibrium and rate constants in complex systems like liquids can be understood at a molecular level, then theoretical chemistry might be your cup of tea. On the other hand, if this all seems way too abstract and mathematical …

The basic ideas of this connection between dynamics and thermo-dynamics are very simple. Consider a potential energy diagram like that shown schematically at the top left of Figure 2.2. If some dividing line is drawn between wells $A$ and $B$, then the equilibrium constant for the interconversion of $A$ to $B$ is simply computed as the fraction of time the simulation resides in region $B$ divided by that for region $A$, and the free energy difference between these states $(\Delta G)$ is proportional to the logarithm of the equilibrium constant:

$$\Delta G = -k_B T \ln \frac{\rho(B)}{\rho(A)}, \tag{2.2}$$

where $k_B$ is the Boltzmann constant and $T$ the absolute temperature. As long as the barrier between the two regions is large relative to $k_B T$, the simulation will spend little time near the barrier, and the calculated free energy difference will be insensitive to the exact location of the dividing line. As an example, you might want to read an early and influential calculation by Bill Jorgensen and coworkers (see Jorgensen 1981 under Further Reading), who studied the gauche to *trans* isomerization of butane in different solvents. Their simulation was long enough to allow the system to cross the barrier between the wells many times so that the statistical probability of being in one or the other could be accurately estimated. For higher barriers, where such transitions may not occur spontaneously on a timescale convenient for simulation, techniques to accelerate transitions can be used, as discussed below.

A straightforward extension of this idea is illustrated at the bottom left of Figure 2.2, where the fraction of time spent in each small region of the reaction coordinate is used to construct a probability profile, $\rho(\delta)$, and from it, the potential of mean force $W$:

$$W = -k_B T \ln \rho(\delta). \tag{2.3}$$

It is generally true that the length of the simulation needs to be greater to obtain statistically converged results for a potential of mean force than for a free energy difference, since fractional residence times are collected for each small interval. Regions of space with small relative probabilities are prone to have poorly determined potentials of mean force.

One way to overcome the limitations of poor sampling in regions of high free energy is to modify the potential energy of the system by adding a biasing or "umbrella" potential that lowers the potential energy of otherwise unfavorable regions of configuration space. Let the "true" potential energy of the system be $U$ and think about adding a biasing potential $U^*(\delta)$. In practice, $U^*$ is often a parabola about a particular point, $U^* = K(\delta - \delta_o)^2$ (see the right side of Figure 2.2), and a complete profile is mapped out via several calculations ("windows") with different values of $\delta_o$. It turns out that the "real" potential of mean force can be easily reconstructed from the biased one; this sort of unphysical manipulation of energies is a common and powerful tool in simulations of biomolecules.

## CONNECTIONS TO CONFORMATIONAL CHANGES IN PROTEINS

These free energy profiles can be used along with ideas from transition state theory to understand a variety of kinetic events (rates of reactions) in biochemistry, at both qualitative and quantitative levels.

According to the transition state model (TST), an estimate of the rate at which systems cross the barrier between wells depends primarily on the barrier height and can be computed as

$$k_{TST} = \frac{1}{2}\left\langle \left|\frac{d\delta}{dt}\right|\right\rangle_{TS} \rho(\delta^*)\Big/ \int \rho(\delta)d\delta. \tag{2.4}$$

Here, the rate constant is the product of the mean velocity of crossing the transition-state surface, $\frac{1}{2}<|d\delta/dt|>_{TS}$, times a probability of being at that surface. This latter quantity depends not only upon the probability $\rho(\delta^*)$ of being at the "transition state" (near the top of the barrier that separates reactants from products) but also upon the "width" of the reactant well: for broader wells, the integral in the denominator of Equation 2.4 will be larger and the rate constant smaller. The mean velocities can be calculated from several theories or directly from dynamical simulations. In addition, corrections to the transition state theory estimate given in Equation 2.4 can also be made from dynamical simulations, but these considerations are beyond the scope of this letter.

## LIGAND BINDING TO MYOGLOBIN

One of the simplest, and best studied, of ligand binding events is that of small diatomic molecules like $O_2$ to the oxygen storage and transport proteins myoglobin and hemoglobin. The origins of its simplicity are straightforward: the ligand has few internal degrees of freedom, and the "chemistry" involves just the reversible formation of a bond to an iron atom of the porphyrin prosthetic group. Furthermore, the ligand binding process can be studied with unmatched time resolution by using short laser pulses to photodissociate bound states followed by optical or resonance Raman probes of the time course of rebinding. Experiments of this sort have provided evidence for the nature of the rebinding process from the subpicosecond regime to seconds and beyond.

Underneath this seeming simplicity, however, lies a wealth of complicated detail about the microscopic events involved. Some of the earliest studies to identify intermediates used low temperatures to slow the rebinding process sufficiently to follow its kinetics after microsecond laser pulses. At very low temperatures (20–140 K), the time course of rebinding is nonexponential, approximating a power law. This can be modeled by assuming a distribution of activation energies arising from microscopically different protein conformations that are "frozen out" at these temperatures. At higher temperatures, conversion among these microstates apparently becomes rapid relative to the rebinding process, so that exponential rebinding is again observed, but with evidence for several kinetic intermediates. In this model, for temperatures above about 140 K (so that interconversion among protein "microstates" is relatively rapid), the time course of rebinding can be represented by a kinetic scheme with one or two intermediates between the bound state and that in which the ligand has diffused far away from the protein. The same general model, with similar kinetic parameters, has been deduced from room-temperature measurements.

The X-ray crystal structure of myoglobin (see Figure 2.3) does not reveal any path by which ligands such as oxygen can move between the heme-binding site and the outside of the protein, so that structural fluctuation must be involved in the entrance and exit of the ligands. Empirical energy function calculations show that the rigid protein would have barriers on the order of 100 kcal/mol; such high barriers would make the transitions infinitely long on a biological timescale. The shortest path for a ligand from the heme pocket to the exterior is between His E7 and Val E11. It is thus of interest to study

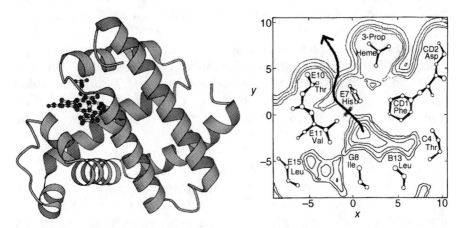

**Figure 2.3.** X-ray structure of myoglobin (left). The helices are shown as ribbons and the heme group as balls and sticks (center left). The oxygen binding pocket is just above the center of the heme group. A slice through the protein, parallel to the iron atom at the center of the heme group, is shown (right). The iron atom is at the origin of the coordinate system, and the plane shown is at z = 3.4 Å. The arrow shows a possible path for the migration of an oxygen molecule from the inside of the protein (at the tail of the arrow) into solution (at the head of the arrow). The transition-state region is shown as crosses between residues E7 and E11.

the energetics of barrier relaxation to determine how a ligand might escape with a reasonable activation energy. Local dihedral rotations of key side chains allow the pocket to be opened at the expense of modest strain in the protein by rigid rotations of the side chains of His E7, Val E11, and Thr E10.

Umbrella-sampling transition-state calculations have been carried out to estimate the energetics of this process. This study began by defining a reaction coordinate $\delta$ as the perpendicular distance from the center of the dioxygen ligand to the plane defined by three atoms of the protein; to explore the path shown in Figure 2.3, these were chosen so that the dividing plane between "inside" and "outside" was perpendicular to the plane of the figure and passed through the side chains of Val E11 and His E7. The simulations then estimated the equilibrium distribution $\rho(\delta)$ by adding a series of umbrella potentials to bias the simulation toward regions of high energy, that is, toward the bottleneck preventing ligand escape from the heme pocket. Since the distributions for the various umbrella potentials overlap, they can be joined to form a continuous function. The free energy barrier to escape is found to be about 6.5 kcal/mol, which corresponds to a rate constant of about $4 \times 10^7$/second, which is within a factor of 4 of the experimental values determined by the room-temperature measurements described above.

Once the location of the barrier in the reaction free energy profile has been determined, a sampling of trajectories may be made by beginning the protein–ligand system at the top of the barrier and by numerically integrating Newton's equations of motion forward and backward in time. This provides a detailed microscopic description of the factors that slow the rate of escape from a diffusion-controlled upper limit down to the values observed experimentally. The simulations can also be carried out as a function of temperature and the results compared to the experiment.

The original work was done 20 years ago, but this has turned out to be an "evergreen" problem, one that has been revisited many times; a good summary is given in the *Proceedings of the National Academy of Sciences of the United States of America* 2008 article listed under Further Reading. There are other ways to get in and out of the protein, and small molecules like oxygen may diffuse to many places and make use of multiple pathways. There are also many computational ways to attack this problem, which is part of the reason that many find this such a satisfying area of research: things are always changing, even as one thinks about some of the most fundamental aspects of protein structure and function.

## WHAT'S NEXT?

OK, I've been long-winded and technical enough! If what I've described sounds intriguing, and if you enjoy math and interacting with computers, give it a try! You can accomplish many things with a personal computer, even on your own, and you don't need access to a "wet" lab. I suggest that you start with a visualization program, like Visual Molecular Dynamics (VMD) (http://www.ks.uiuc.edu/Research/vmd/) or Chimera (http://www.cgl.ucsf.edu/chimera/). You can download these onto a PC or Mac, and each has excellent tutorials to get you started. To go further, look for undergraduate research opportunities, not only in the obvious theoretical groups in your department but in experimental labs as well. The techniques described here are widely used, and not just by theoreticians. You'll learn a lot, even if you don't make it your life's work.

Best regards,

Dave Case

## FURTHER READING

Case, D. A.; Karplus, M. Dynamics of ligand binding to heme proteins. *Journal of Molecular Biology* 1979, *132*, 343–368.

Harvey, S.; McCammon, J. A. *Dynamics of Proteins and Nucleic Acids*, Cambridge University Press, Cambridge, 1987.

Jorgensen, W. L.; Binning, R. C. Jr; Bigot, B. Structures and properties of organic liquids: *n*-butane and 1,2-dichloroethane and their conformational equilibria. *Journal of the American Chemical Society* 1981, *103*, 4393–4399.

Ruscio, J. Z.; Kumar, D.; Shukla, M.; Prisant, M. G.; Murali, T. M.; Onufriev, A. V. Atomic level computational identification of ligand migration pathways between solvent and binding site in myoglobin. *Proceedings of the National Academy of Sciences of the United States of America* 2008, *105*, 9204–9209.

# 3

# The Purple Planet: A Short Tour of Porphyrins and Related Macrocycles

**Abhik Ghosh**

*University of Tromsø, Norway*

Born in Kolkata, India, Abhik Ghosh received his PhD in 1992 from the University of Minnesota, working with Regents' Professor Paul Gassman. After postdoctoral stints with Professors Larry Que and David Bocian, he crossed the Atlantic in 1996, taking up a faculty position at the University of Tromsø, Norway, where he is now professor of chemistry and a principal scientist at the Center for Theoretical and Computational Chemistry. From 1997 to 2004, he was a senior fellow at the San Diego Supercomputer Center. Subsequently, he has been a visiting professor at The University of Auckland, New Zealand, on multiple occasions. His research interests span bioinorganic chemistry, materials chemistry, and computational chemistry. He has published about 125 research papers and edited *The Smallest Biomolecules*, a bestselling book on diatomics and their interactions with heme proteins. Outside of work, he enjoys travel, hiking, bird-watching, and all manner of natural history, especially in the company of his son Avroneel.

Hi Angela!

Good to hear from you again! It was great having you here for a semester. It was rather gutsy, if you don't mind my saying so, to leave sunny Southern California in January and head for the world's northernmost university as an exchange student. Well, we have your Uncle Carl (*Editor's note*: This refers to Professor Carl Wamser, who has written Chapter 16) to thank for helping you make up your mind. I am glad it all worked out so well, both scientifically and personally. My group members still talk about the great times they shared with you. I hear you are in touch with them via Facebook on a regular basis. My busy schedule and travel commitments didn't really allow me to get to know you outside of work, but I'm glad we managed to have a couple of barbecues. Those late evenings by the fjord were lovely!

You seemed to enjoy making porphyrins, corroles, and their metal complexes in our lab in Tromsø, and I'm flattered that you are considering the field as a potential career direction. Let me give you an introduction to the field, focusing on the basics, but also pointing out some exciting research directions. Porphyrins are beautiful, literally! I still remember my own first experience with them some 20 years ago: how filtering a flask of disgusting black sludge left a mound of shining purple crystals of tetraphenylporphyrin! It was magic and I have never broken free of that spell. Well, beauty can be a surprisingly important consideration in science. Of course there are also many *rational* reasons for taking an interest in porphyrins and their structural variants. Below I'll give you a small sampler of the world of porphyrins and related compounds—the purple planet, as I sometimes call it.

## PORPHYRINS, PORPHYRINS EVERYWHERE

Porphyrins are everywhere, as far as the living world is concerned. Their best-known forms (Figure 3.1) are heme, the red pigment that gives hemoglobin its red color, and chlorophyll, a reduced porphyrin, which is responsible for the green color of plants. The basic skeleton

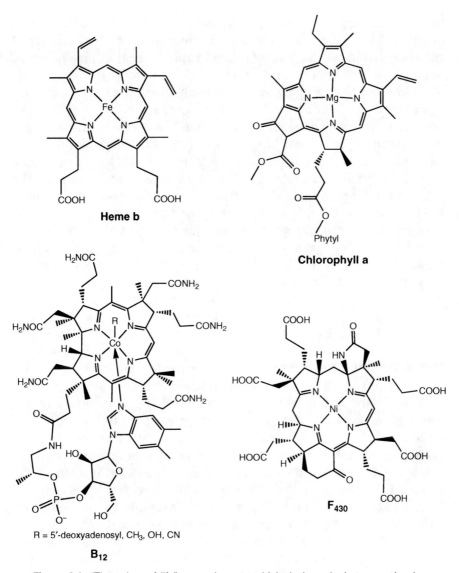

**Figure 3.1.** *"The colors of life": some important biological porphyrin-type molecules.*

of these molecules probably arose by some kind of self-assembly-like process from smaller molecules, not unlike some of the one-pot porphyrin and corrole syntheses you carried out in Tromsø. What fascinates me is how hemes and their cousins became such universal and critical cogs in the machinery of life. The fact that they occur in all the domains of life—*Archaea*, *Bacteria*, and *Eukarya*—indicates that they became a part of biology very early on, at or very close to the origin of life. Their role increased enormously as the availability of chlorophyll led to the rise of photosynthesis and, gradually, to an oxygen-rich atmosphere and aerobic life. Today, porphyrin-type cofactors are recognized as playing a stunningly diverse set of biochemical roles. Hemoglobin and myoglobin are surely familiar to you as oxygen carriers in much of the animal kingdom. What you might not have heard about is that these proteins also interact with NO, a diatomic molecule increasingly recognized as a ubiquitous biological signal ("the smallest hormone"). There are numerous other examples. The heme-containing protein cytochrome $P_{450}$ uses molecular $O_2$ to oxidize C–H bonds to C–OH, a notoriously difficult process for organic reagents or "purely organic" enzymes. The heme proteins cytochromes b and c are key links in the respiratory electron transfer chain. Observe from Figure 3.1 the structure of the $B_{12}$ cofactors, which, while broadly similar to heme (i.e., an iron porphyrin), is also clearly different: $B_{12}$ is much more saturated than the fully conjugated and aromatic porphyrins and also contains cobalt instead of iron. Also shown in Figure 3.1 is $F_{430}$, a highly saturated nickel-porphyrin-like cofactor, which occurs as a part of the enzyme methylcoenzyme M reductase. Present in methanogenic *Archaea* (which live in anaerobic environments, such as submerged paddies and cow stomachs), this enzyme catalyzes the last step of biological methane production. I could go on and on, but you get the point: porphyrins are everywhere, and the best part is that new heme-containing proteins with surprising functions are still being discovered at a regular pace.

I won't focus here on the architectural details of heme proteins. Instead I'll zoom in on the heme itself and try to give you a sense of its functions. Cytochrome $P_{450}$ is a good example for this purpose. After the Fe(II) center in the heme binds molecular $O_2$, it undergoes a few more reactions (shown in Figure 3.2, but which I won't describe in any detail), culminating in the generation of a highly reactive intermediate called "compound I." This is a remarkable species with a formally $Fe^{+5}$=O center. If you look at the section on iron in your sophomore inorganic text, you'll probably find a fairly plain statement that it occurs in the +2 and +3 oxidation states in its compounds. Perfectly true, but that also tells you that an $Fe^{+5}$ species is a pretty wild concept! Well, let me backtrack a tiny bit: writing $Fe^{+5}$=O is not 100% kosher; compound I is probably best described as an $Fe^{+4}$=O center coupled to a ligand radical. Whatever the detailed description, the key point is that the overall oxidation state of the heme has been jacked up a phenomenal two notches above $Fe^{+3}$,

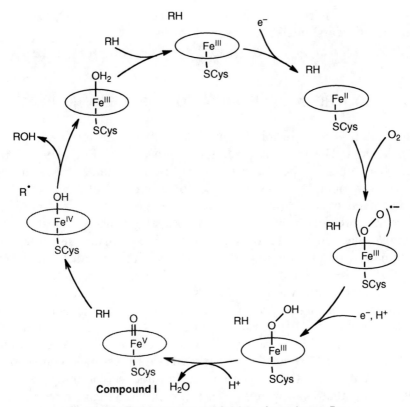

**Figure 3.2.** *The consensus mechanism of cytochrome $P_{450}$.*

giving compound I the muscle to tear off a hydrogen atom from one of those famously strong C–H bonds.

High-valent transition metal intermediates, for me, symbolize the amazing give and take between chemistry and biology. Of course, chemistry underpins all of biology; in a sense, biology *is* chemistry, or as someone said more colorfully, biology is chemistry that *works*! What is less appreciated is how much fundamentally new chemistry we learn from biological systems. Were it not for cytochrome $P_{450}$ and other iron-containing enzymes, for example, our knowledge of fundamental iron chemistry would be far more limited than it is. In the same spirit, biological methane production involves a good deal of funky nickel chemistry, based on the $F_{430}$ cofactor I mentioned. This chemistry involves both the unusual +1 and +3 states of nickel, which normally occurs as +2 ions. Similarly, the biochemistry of $B_{12}$ cofactors involves remarkable cobalt chemistry. For all the weird molecules dreamed up and synthesized by inorganic chemists, they do learn a thing or two from Mother Nature, who, after all, has had a few billion years' head start over all of us.

## FROM SOUTHERN CALIFORNIA TO NORTHERN NORWAY

Let me tell you a bit about how I developed as a research chemist in academia. The years immediately following my PhD (1993–1994) were bad years for faculty recruitment in the United States, so I kept my options open and was prepared to relocate to other parts of the world, if necessary. Add to this scenario the fact that my graduate advisor, Professor Paul Gassman, a very distinguished physical organic chemist, died unexpectedly, and I lost a father figure and advocate. Fortunately, I had collaborated extensively with a Swedish theoretical chemist at the University of Minnesota, Professor Jan Almlöf, during my graduate years, and one day he asked me whether I had ever considered working in Europe. He mentioned that the University of Tromsø in Norway had a vacancy and that they were interested in a person like me, a theoretically inclined experimentalist. After a quick look at the map (Tromsø is at nearly 70°N), I must have come across as skeptical. Jan assured me that the latitude was nothing to worry about and that the winters were far milder than they were in Minneapolis. As it happened, Jan had learned about this opening from his friend, Professor Odd Gropen, from the University of Tromsø, who at the time was spending his sabbatical year in Minneapolis. In a couple of weeks, at a party at Odd's place, both Jan and Odd tried to sell me on Tromsø and my wife, an adventurous spirit, decided to give Tromsø a try, if an offer were to come our way. I duly applied to Tromsø and waited. The wheels of academic hiring turn slowly in Norway so I saw no problems with moving to a postdoc in California before a possible move to Norway. Besides, in January 1994, I had been on my first visit to California in connection with the Metals in Biology Gordon Conference. Coming from frozen Minneapolis, I was blown away the moment I stepped out of the Los Angeles airport—the soft breeze, the greenery, the flowers, and girls and guys in shorts! Well, I decided that I *had* to do a postdoc in California before I took up a position in Norway!

In May 1995, I was invited to Norway for an interview, which went well. Odd and his wife, who later became good friends of mine, were the most gracious hosts imaginable: we lingered over roast ptarmigan (snow grouse) as the snowy peaks in the picture window turned pink in the midnight sun. It wasn't hard being sold on Tromsø, especially since the weather was gorgeous all through my 4-day visit, which, as you know, is not something you always take for granted! Summer turned to fall, and one day, an offer from Tromsø turned up in our mailbox in California. I hadn't really done a proper job search in the United States until then, and I agonized over whether I was taking the easy way out by accepting a position in faraway Norway. We discussed the matter over dinner at our favorite Chinese restaurant in Riverside, and I decided to accept the offer. It was a permanent position and, although Tromsø couldn't offer all the facilities available at a large American university, it was a perfectly reasonable place to live and pursue a research career. Besides, the outdoors, always an important consideration for me, was amazing. So, I accepted the offer but asked for (and was granted) a slight deferral so our son, born later that fall, didn't have to travel while still a newborn.

I hope I haven't bored you witless with my life story, but I really want to make a point. For most scientists, life is not a predictable, linear path: straight-A student in high school → graduate research with a famous professor → postdoc with a Nobel laureate or equivalent → 5–10 job offers to choose from. My life's path was curvier and had its share of setbacks. I already mentioned my PhD advisor's death; Jan too passed away before I started my position in Tromsø. Throw in other challenges like short stretches of unemployment and broken relationships—the all too common experiences of young people—and it can seem like a minor miracle that so many people do end up with successful academic careers in science. Academic research positions in science are relatively few, so your dream job is hardly guaranteed once you have completed your PhD or even a postdoc. Science is certainly a harder career path compared with law, medicine, finance, computers, and so on, where the job market is bigger and hungrier for fresh graduates. But then, a career in the arts is tough as well. I have a number of artist friends and at some point, they all faced far more uncertain futures than I ever did. I am glad that I hung in there and was on the lookout for lucky breaks. Today I wouldn't trade my job for anyone else's!

## PORPHYRIN ANALOGUES

I came to Tromsø in the spring of 1996. I say spring because it was April, but the snow was still thick on the ground. Fortunately, the weather was gorgeous, as it often is in spring up here, so we were in high spirits. I already had some independent, ongoing projects: I had taught myself about density functional theory (DFT), which is essentially an alternative formulation of quantum chemistry, largely as a hobby during my postdoctoral years, and was using it to explore the electronic structures of a variety of transition metal complexes. DFT was still very new to inorganic chemists and transition metals were a wide-open field for exploration. But I'll come back to theoretical chemistry later. My prime concern that year was to get an experimental program started. That was easier said than done. I had an empty lab with substandard hoods and no students, and whoever had heard of setup funds in Norway in those days? Odd Gropen gave me some good tips about grant writing in Norway and before the end of my first year, I had quite a few students, both doctoral and master's. The labs were still a headache: if Schlenck lines (remember the glass contraptions you occasionally used for handling air-sensitive compounds?) were hard to get made in our rudimentary glass shop, glove boxes were a distant dream. Fortunately, things were afoot in the porphyrin field, which I was excited to explore and which didn't demand too much by way of lab resources.

What was happening was that porphyrin chemists were isolating a variety of macrocycles from the black gunk left behind at the end of a regular one-pot porphyrin synthesis. So you obtain not only porphyrins but also *N*-confused porphyrins, corroles, sapphyrins, and numerous expanded porphyrins—in short, a zoo (Figure 3.3). Some of these ring systems, such as corroles and sapphyrins (but not N-confused porphyrins), were previously known, but only as end products of tedious, multistep syntheses. Now, we could access them with almost ridiculous ease. Importantly, it was possible to tinker with the conditions so we could selectively obtain one macrocycle as opposed to the entire zoo. We had little understanding of the kinetics and thermodynamics of the paths whereby the pyrrole and aldehyde monomers and their various oligomers self-assembled into these wonderful rings, but that didn't bother us. We were overjoyed that these ligands essentially made themselves and wanted nothing more than to explore their metal-binding ability and other properties.

My good Israeli friend Zeev Gross had just reported that a corrole could be obtained from pyrrole and pentafluorobenzaldehyde under solvent-free conditions. Although he emphasized the importance of an electron-deficient aldehyde, we found that his reaction led to corroles for electron-rich aldehydes as well. A PhD student of mine, Erik Steene, showed that the reaction worked not only for pyrrole but also for 3,4-difluoropyrrole (which, unfortunately, is a pain to

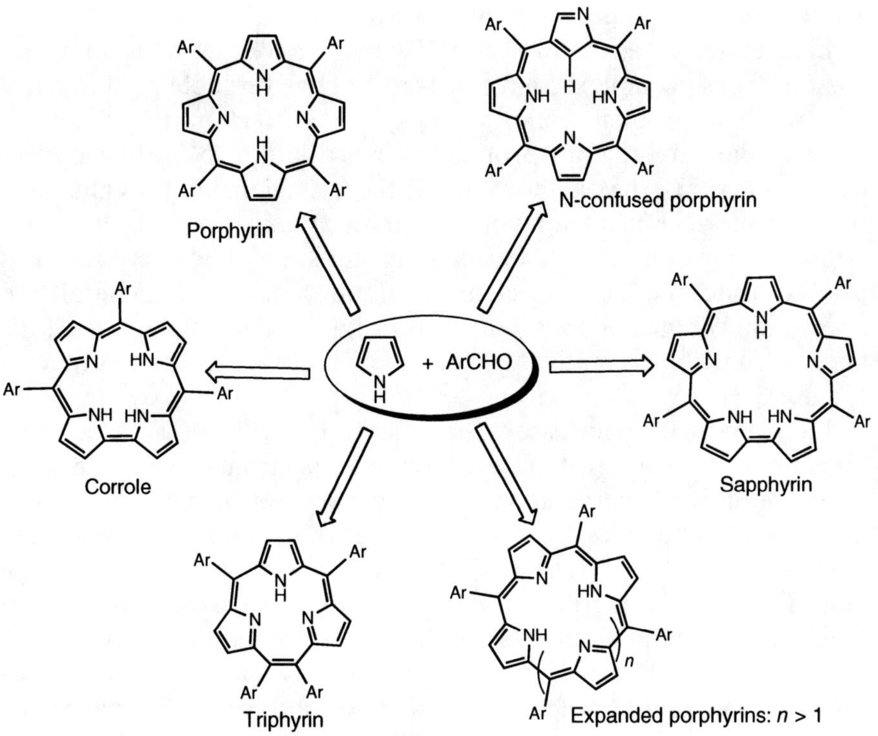

Porphyrin

N-confused porphyrin

Corrole

Sapphyrin

+ ArCHO

Triphyrin

Ar    Expanded porphyrins: *n* > 1

**Figure 3.3.** *Pyrrole–aldehyde condensations as a source of macrocyclic ligands.*

make), leading to corroles fully fluorinated at the pyrrole β positions. For a while, it seemed as though substituted corroles could be obtained as simply and reliably as porphyrins. We were surprised and disappointed when our trusty corrole synthesis didn't work well for other substituted pyrroles such as 3,4-dichloropyrrole and 3,4-diethylpyrrole. We haven't studied the details yet, but it seems that one-pot corrole synthesis is quite sensitive to steric effects, especially of groups at the pyrrole β positions. I am looking forward to confirming this hypothesis, a worthwhile endeavor in my opinion, considering these simple preps in one step give rather complex and useful ligands.

## A SPECTROSCOPIST'S GOLD MINE

As a result of these studies, we quickly assembled a large supply of corrole ligands with systematically varying peripheral substituents.

These we converted straightforwardly to a variety of metal complexes—manganese, iron, cobalt, copper, silver, molybdenum, you name it. These complexes have proved to be a veritable gold mine of spectroscopic problems, and, to my great satisfaction, the results of these studies are really a wonderful contribution to inorganic electronic structure. Angela, the value of these fundamental insights may be a bit more difficult for you to appreciate than, for instance, our efforts to develop corrole-based solar cells and liquid crystals, but bear with me: the basics of chemistry (and of science in general) are always important and, without a good knowledge of the basics, it's hard to contribute creatively to complex, real-life scientific problems (Figure 3.4).

Talking about spectroscopy, let's start by defining an electronic absorption spectrum (often called a uv-vis spectrum or simply a spectrum). Well, it's a curve showing the absorption of light by a given material as a function of the wavelength (I guess you knew that, having gotten at least ten uv-vis spectra every day you worked in my lab). The absorption of light (i.e., photons) excites electrons from their stable "home" orbitals into (often previously unoccupied) higher-energy orbitals. The peaks in a uv-vis spectrum thus correspond to the energies of the excited states and, with a bit of experience and analysis, they tell you a fair amount about the electronic

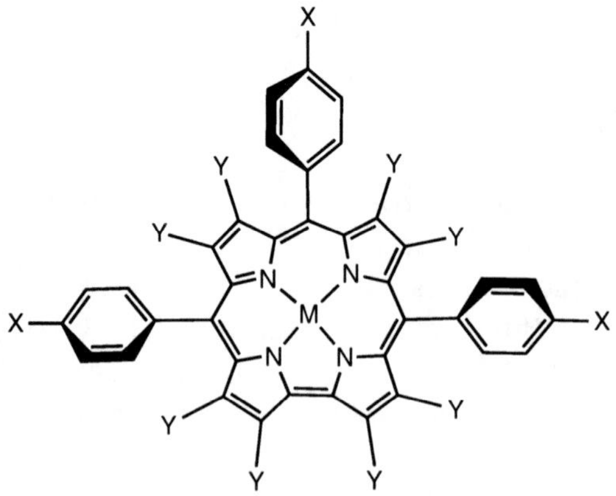

$$M = H_3, CrO, Mn, MnCl, FeCl, FePh, Co(PPh_3), Cu, MoO, Ag$$
$$X = CF_3, NO_2, CO_2Me, COOH, H, F, CH_3, OCH_3$$
$$Y = CF_3, Br, Cl, F, H, CH_3$$

**Figure 3.4.** *Metallocorroles synthesized in our lab. Many, but not all, combinations of M, X, and Y have been synthesized.*

structure of the molecule in question. For a porphyrin without a transition metal at the center, these energies typically correspond to excitations of the aromatic π-system, the large circular network of relatively loosely bound electrons. In the presence of a coordinated transition metal ion, with all its d electrons, the spectra get a bit more complicated but also a good deal more fun to analyze.

Transition metal corroles are extra special because they are typically high-valent. You probably remember our dozens and dozens of vials containing formally Cr(V), Fe(IV), Cu(III), and Ag(III) complexes; it's a pretty incredible collection when you consider the rarity of these oxidation states. Recall that for porphyrins, such oxidation states occur, if at all, only in the form of highly reactive intermediates. Now have a look at the uv-vis spectra of a family of copper corroles with varying *para*-substituents on the corrole *meso* phenyl groups (Figure 3.5). You'll see that the spectra, especially the biggest peak (the so-called Soret band), shift to conspicuously longer wavelengths as the *meso*-substituent becomes increasingly electron donating. This behavior is in sharp contrast to that observed for analogous porphyrin derivatives, whose spectra remain pretty much on top of one another. The simplest explanation for this effect is that the corrole Soret band doesn't just involve the macrocycle π-system (as it does for a "simple" metal–porphyrin complex); instead, it has what is called a charge-transfer component, that is, an excitation of an electron from the peripheral phenyl groups into the high-valent core

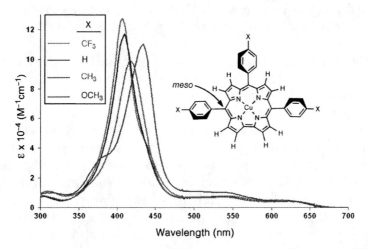

**Figure 3.5.** *Electronic absorption spectra of a series of copper corroles with systematically varying substituents.*

of the molecule. Theoretical studies suggest that this simple explanation is indeed correct for copper corroles. I derive a certain joy from being associated with what some might view as a somewhat academic discovery: the shifting Soret bands of the copper corroles provide some of the most beautiful examples of charge-transfer transitions, the likes of which are essentially unknown for porphyrins, and that's something worth taking pride in.

## THE END OF INNOCENCE

I already wrote that many metallocorroles are high-valent, but that's far from the whole story. The trianionic corrole$^{3-}$ ligand is easily oxidized, and the corrole ligand in many metallocorroles is best described as a corrole$^{\cdot 2-}$ radical. Inorganic chemists describe such ligands as noninnocent, a fanciful but well-established term meaning that the ligand cannot be accurately described with a simple Lewis structure. Professor F. Ann Walker of the University of Arizona, Tucson, was the first to provide concrete evidence based on nuclear magnetic resonance (NMR) spectroscopy that certain iron corroles, specifically Fe(corrole)Cl complexes, were noninnocent; that is, they were not quite Fe$^{IV}$(corrole$^{3-}$)Cl but were better described as Fe$^{III}$(corrole$^{\cdot 2-}$)Cl. It was a fascinating description that I was able to quickly confirm with DFT calculations, albeit somewhat qualitatively. Unfortunately, not everyone accepted this picture of metallocorrole electronic structure and a controversy ensued. That was tiresome, even painful; yet, looking back, it was not altogether a negative episode in our corrole research.

I still remember an American Chemical Society (ACS) meeting in New York City, where Ann and I sat together in a forty-seventh-floor café discussing corrole research. Ann seemed completely oblivious to the panoramic view around us but was clearly perturbed by what we both felt was a needless controversy. Back in my hotel room, I couldn't help smiling to myself: Here were some of the most distinguished inorganic chemists I knew, agonizing over whether half an electron was on the iron or on the ligand! Who cared about something that arcane? I remembered something written by Peter Medawar—that scientific controversies excite as much passion as they do precisely because they tend to revolve around trivial issues; they might not involve large sums of money, but they involve ideas and concepts that scientists hold dear and hate to abandon.

That said, let me backtrack a bit and say that more than a few people do care about the nature of high-valent iron species. I have already told you that such species occur as intermediates in some important enzymes such as cytochrome $P_{450}$. So, although it's a bit exotic, there's certainly no shortage of interest in the problem we were studying. As there were no more obvious experimental approaches to the problem, I longed for a theoretical approach, one much more definitive than DFT, which would resolve the controversy once and for all. Ultimately, I did manage to do that, but it took years of thinking, planning, and hard work.

Very recently, an unexpected line of evidence brought home to me the importance of ligand noninnocence in metallocorrole chemistry. A student of mine (Abraham Alemayehu) managed to obtain X-ray crystal structures of two different copper corroles, and both exhibited oddly nonplanar saddle-shaped conformations; recall that the corrole ring system is aromatic (it's in a sense a bigger version of benzene) and metallocorroles, by and large, are not expected to be nonplanar. Moreover, because they lack one of the *meso* carbon bridges of porphyrins, corroles are stiffer than porphyrins with respect to out-of-plane deformation. It seemed unlikely that the odd saddle-shaped copper corrole structures could arise from weak crystal packing forces; instead, the structures clearly suggested a definite electronic effect at work. Using DFT calculations, we mapped out the energy necessary for buckling different metallocorroles out of planarity and into a saddle-shaped conformation. The resulting energy curves clearly showed that copper corroles are *inherently* saddle-shaped, whereas most other metallocorroles are planar. As suspected, the inherent saddling of copper corroles results from a specific copper–corrole orbital interaction. I won't discuss details, but the upshot of our analysis is that copper corroles aren't the straightforward $d^8$ Cu(III) complexes we thought they were; even in their diamagnetic ground states, they are best viewed as coupled $Cu^{II}$–corrole$^{•2-}$ entities. I briefly mentioned these findings to Ann at the last ACS meeting in Salt Lake City; she seemed happy that her early insights were echoed in yet another context.

## THE ROLE OF THEORY

I know you haven't had a formal introduction to quantum chemistry, but you've probably learned a few things about molecular orbitals.

You probably also know that perhaps the most important tool at our disposal today for theoretical studies of structure and bonding is DFT. DFT is based on a fundamental theorem of quantum mechanics, which states that the energy and other properties of an electronic system in its ground state is completely determined by the electron density. DFT is a way of solving the Schrödinger equation that emphasizes the electron density and is philosophically distinct from wavefunction-based ab initio methods. The great advantages of DFT are that it's relatively quick and it gives generally good results for transition metals. Today, as far as most organic and inorganic chemists are concerned, quantum chemistry and DFT are often synonymous (perhaps to the shock of some purists), and traditional ab initio methods are far less popular than they used to be.

My key collaborator during my PhD years, Jan Almlöf, was a superb ab initio theorist. Working with him initially and then on my own, I developed a good sense of when one can get by with DFT and the rather rare occasions when one cannot. I wrote above that I particularly treasure problems that appear to defy theoretical modeling (read DFT). A number of these problems are satisfactorily resolved by high-level ab initio methods. The spin state energetics of a transition metal complex is a good example of such a problem. DFT is often unreliable for calculating the energy difference between the high- and low-spin states of a transition metal complex, which are characterized by the maximum and minimum possible numbers of unpaired electrons. Fortunately, ab initio methods can tackle this problem. The metallocorroles, however, present a more complex problem. When dealing with FeCl corrole, for example, the ideal method must not only provide a good description of iron spin states but also correctly describe metal-centered versus ligand-centered oxidation, that is, whether the corrole ligand is 3– or ·2–. These requirements are met by the so-called multiconfigurational ab initio methods.

High-level ab initio calculations are no child's play, and a low-symmetry system the size of a porphyrin or corrole is at the limit of what is doable today. To increase my chances of success, I teamed up with two leading practitioners (indeed architects) of modern multiconfigurational quantum chemistry, Peter Taylor of the San Diego Supercomputer Center (by the way, Pete was my key San Diego connection) and Björn Roos of Lund University, Sweden. Despite the formidable gathering of expertise, the FeCl corrole calculations tried our skills and ingenuity to their limits. After years of hard work and false starts, we obtained clear, conclusive results, and our early picture of a noninnocent electronic structure, $Fe^{III}(corrole^{.2-})Cl$, came out intact.

An important advantage of the multiconfigurational theory is that it allows us to characterize not only the ground state of a system but also several excited states. My hope was that these calculations would not only confirm an $Fe^{III}(corrole^{.2-})Cl$ ground state but also show that a true $Fe^{IV}$ state was quite a bit higher in energy. That would effectively put the controversy to rest. That, however, wasn't exactly how things turned out. What the calculations showed was that *all* low-energy states of Fe(corrole)Cl were best described as $Fe^{III}(corrole^{.2-})Cl$! They only differed with respect to the spin state of the Fe(III) center and the nature of the coupling—parallel or antiparallel—with the corrole radical; the Fe(IV) state was nowhere to be found. So finally, it was all sorted out; it felt like the end of an era!

## FROM FUNDAMENTALS TO APPLICATIONS

Well, Angela, I hope I haven't lost you completely amid the details of the metallocorrole electronic structure. I've tried to give you a flavor of how one goes about doing such analyses using a combination of spectroscopy, X-ray structure determination, and quantum chemical calculations, as well as other techniques that I haven't had a chance to describe. These approaches continue to be useful, even as we have moved away somewhat from exploring fundamental aspects of porphyrins and corroles to developing them as high-tech functional materials.

A major current effort in our lab is to develop corrole-based dye-sensitized solar cells (DSSCs), which you've surely heard about from your Uncle Carl, with whom I've had a wonderful collaboration these last few years. DSSCs are ultimately modeled after photosynthetic light-harvesting arrays, albeit rather crudely. Like others in the business, we prefer to use synthetic porphyrins (or, in our case, corroles) instead of chlorophylls because the synthetic stuff is considerably more rugged and processable. DSSCs in essence consist of an organic dye bonded to titania or another nanostructured semiconductor, and, in principle, they are far cheaper to manufacture than conventional solar cells based on crystalline silicon. Assuming they live up to their promise, they could revolutionize solar energy conversion.

Another potential application of our corroles is in the area of photodynamic therapy (PDT) of cancer and certain other diseases such as psoriasis and macular degeneration. The treatment involves administration of a porphyrin or other dye to the affected tissue and irradiation with red or near-infrared (IR) radiation. Other visible lights wouldn't do because the human body is largely opaque to the shorter wavelengths. The light excites the porphyrin to an excited singlet state, from where it "intersystem-decays" to a relatively long-lived excited triplet state. Now recall that molecular $O_2$, which is generally around in tissues, is a ground-state triplet (i.e., it has two unpaired electrons with parallel spins). The triplet porphyrin dumps its excess energy to $O_2$, which, as a result, is excited to its singlet state. Now singlet $O_2$ reacts aggressively with organic matter, which in this case is a tumor or other diseased tissue. Damaged by singlet $O_2$, the attacked cells perish, generally via apoptosis. Using our systematic knowledge of substituent effects, we have developed a number of corroles with strong absorption in the red and near IR. I wish I could write you that we've hit upon an amazing new sensitizer for PDT, but testing is underway and we are very much looking forward to the outcome.

Red and near-IR absorbers are also of great interest for biomedical imaging, especially a type of imaging based on "two-photon absorption" (TPA, an intriguing phenomenon consisting of the simultaneous absorption of two photons by a molecule). Normally, a very low-intensity process (the uv-vis spectra discussed above all involve single-photon absorption), TPA can be amazingly strong for dimeric or oligomeric porphyrin derivatives as well as for certain weakly bound supramolecular assemblies. Once again, regular visible absorbers aren't that useful because of the strong absorption of most visible wavelengths by human tissues. A great advantage of a TPA-based imaging agent is that it can be located very precisely in a tissue or other biological material because the surrounding tissue will have essentially zero background absorption. Once again, we are very excited about the performance of certain of our corroles as two-photon absorbers.

By the way, in case the above projects come across as a laundry list and you wonder whether there is some kind of common thread across them, the answer is "yes." It all flows from our interest in the fundamentals of porphyrins and corroles, their electronic structure and spectra, and chemical reactivity. We are simply following these interests to their logical conclusion, that is, some useful applications. A case in point is your good friend Adam Chamberlin, who is doing his postdoctoral project on large-scale supercomputer calculations on porphyrin and corrole TPA and trying to formulate simple rules for designing strong two-photon absorbers. He is a quintessential theorist, but I wouldn't be surprised if in 2 years he were named coinventor of a hot new TPA imaging agent. Basic and applied research go hand in hand in our lab; I think that's incredibly stimulating. It's important not to look down on one or the other.

Well, I hope you enjoyed the little tour of the purple planet! I certainly look forward to seeing you again before long, which probably should be at the next ACS meeting in San Diego. And if you want to visit me again here in Norway, you are of course welcome. Good luck with your sophomore year.

Abhik

## FURTHER READING

Ghosh, A. A perspective of pyrrole–aldehyde condensations as versatile self-assembly processes (mini review). *Angewandte Chemie International Edition* 2004, *43*, 1918–1931.

Ghosh, A. (ed.). *The Smallest Biomolecules: Diatomics and Their Interactions with Heme Proteins*, Elsevier, Amsterdam, 2008.

Milgrom, L. R. *The Colours of Life: An Introduction to the Chemistry of Porphyrins and Related Compounds*, Oxford University Press, Oxford, 1997.

# 4

# *Anesthesia: Don't Forget Your Chemistry*

**Jonathan L. Sessler**

*The University of Texas at Austin*

**Daniel I. Sessler**

*The Cleveland Clinic*

Professor Jonathan L. Sessler received a BS degree in chemistry in 1977 from the University of California, Berkeley. He obtained a PhD in organic chemistry from Stanford University in 1982 working with Professor James P. Collman. He was a National Science Foundation-Centre National de la Recherche Scientifique (NSF-CNRS) and National Science Foundation-North Atlantic Treaty Organization (NSF-NATO) postdoctoral fellow first with Professor Jean-Marie Lehn in Strasbourg, France, and then with Professor Iwao Tabushi in Kyoto, Japan. In September 1984, he accepted a position as assistant professor of chemistry at the University of Texas at Austin, where he is currently the Roland K. Pettit Chair. Professor Sessler has authored or coauthored over 500 research publications and has written two books (with Dr. Steven J. Weghorn and Drs. Philip A. Gale and Won-Seob Cho, respectively). He is also an inventor on more than 75 issued U.S. patents. Professor Sessler has cofounded

*Letters to a Young Chemist*, First Edition. Edited by Abhik Ghosh.
© 2011 John Wiley & Sons, Inc. Published 2011 by John Wiley & Sons, Inc.

two companies—Pharmacyclics, Inc. and Anionics, Inc.—and has served as co-organizer of several international conferences on porphyrin and supramolecular and macrocyclic chemistry. In addition to English, he speaks French, Spanish, German, and Hebrew, can get by in Japanese and Italian and is studying Korean.

Dr. Daniel I. Sessler attended medical school at Columbia University and subsequently completed pediatric and anesthesiology residencies at the University of California, Los Angeles (UCLA). He is currently professor and chair of the Department of Outcomes Research at the Cleveland Clinic. The Outcomes Research Consortium, which Dr. Sessler founded, includes 100 investigators in 10 countries. Outcomes Research currently coordinates more than a hundred studies, including a dozen large, multicenter outcome trials. Dr. Sessler has published a book on therapeutic hypothermia and more than 425 full research papers. Dr. Sessler has trained 70 research fellows, four of whom subsequently chaired anesthesia departments. Among his awards may be mentioned a Fulbright Fellowship and the 2002 American Society of Anesthesiology Excellence in Research prize.

Hi Angela,

I am just back from a visit to my brother, Daniel I. Sessler, MD. He is a professor of anesthesia and chair of the Department of Outcomes Research at the Cleveland Clinic in Ohio. You don't remember him, but he remembers you. Years ago, when you were 7 years old, you may recall being rushed to UCLA for an emergency appendectomy. If you think about it at all, you probably remember the loving attention surrounding your recovery and, looking back, you probably think of the surgery as scary but basically routine. In fact, you almost died.

My brother was an anesthesiology resident at UCLA at the time and he remembers your case because you experienced a rare and serious complication. You are with us today because he quickly and accurately diagnosed your condition and instituted the appropriate treatment. Since you are currently taking organic chemistry and starting to consider career options, I am sure you will find it interesting to reflect and appreciate how much this diagnosis and, indeed, so much of modern anesthesia rely on a detailed understanding of chemistry. However, there are also many aspects of anesthesia that are far from well understood. These are opportunities for future research. I hope you will consider them as you continue to develop your skills as a chemist. Meanwhile, let's discuss what happened during your anesthesia, which is as good a way as any to summarize the state of the art in this fascinating field. It will also allow my brother to explain why, thanks to chemistry, you're still alive.

When you came into surgery, you didn't know it, but you were about to embark on nothing less than a great chemical adventure. The human body is arguably the most complex chemical system known. Almost all drugs, including anesthetics, work by modulating the chemistry of our bodies. In many cases, the underlying mechanisms are well understood and involve a defined activation or inhibition of a key biological process; however, in many other cases, they are not. One of the great challenges in drug development, and where chemists can have an impact, is in understanding how known drugs work, which in turn can provide the basis for the design of even better drugs. (For the remainder of this letter, we will use the conventional definition of a drug, any substance that, when injected into a rat, produces a scientific paper!) One important approach involves understanding how specific changes in structure relate to key parameters, such as potency, toxicity, clearance rates, and so on, since this often provides a basis for optimizing a given class of agents for clinical use. This principle is well illustrated by the anesthetics you received as a 7-year-old.

Anesthetics can be broadly divided into local and general anesthetics, and the category of general anesthetics can be subdivided into intravenous and inhalational anesthetics. The components of general anesthesia are *hypnosis* (unconsciousness), *analgesia* (lack of pain), control of reflexes, such as heart rate and blood pressure, and muscle relaxation (to prevent patients from getting up and going home in the middle of surgery). You were given a drug from each anesthetic class, each for a specific purpose. This care and design is typical of a modern medical treatment and is a key to making surgery such a success. However, the choice and administration of anesthetics is extremely tricky. Normally, people worry about the surgery; however, the risks associated with anesthesia often exceed those associated with surgery. This is especially true for so-called minor surgery. As your own story illustrates, there is no such thing as minor anesthesia.

## LOCAL ANESTHETICS

The first anesthetic you were given, lidocaine, was injected into the back of your left hand. It was given to numb the site where a few

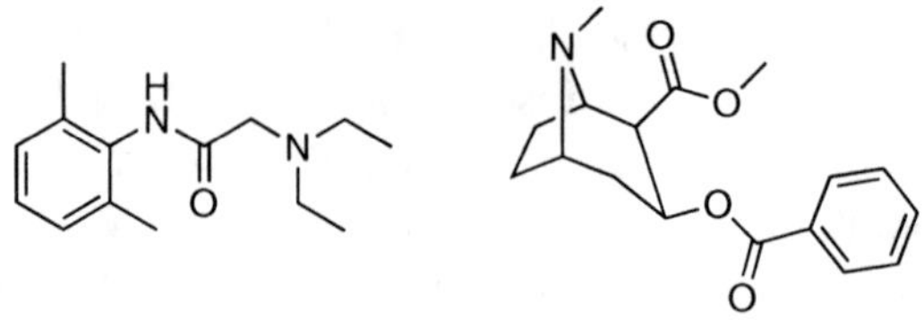

**Figure 4.1.** *Structures of lidocaine (left) and cocaine (right). These compounds are administered and used as local anesthetics in the form of their hydrochloride salts. In these protonated forms, they act as sodium cation mimics and block ion channels critical to the propagation of nerve signals.*

minutes later an intravenous catheter was to be placed. Lidocaine is a simple amine compound, something you could prepare easily in an organic laboratory. It is a synthetic mimic of complex natural nitrogen-containing alkaloids such as cocaine (see Figure 4.1 for structures of lidocaine and cocaine). Interestingly, cocaine, whose analgesic properties were recognized by the Incas centuries ago, was the first compound to be used as a local anesthetic by Western medical practitioners. For a time, it was the only known anesthetic. Cocaine is a strong vasoconstrictor but is better known for its psychoactive properties (i.e., it modifies neurological function, and thus mood and behavior). It is therefore a controlled substance and is now rarely used medically, although it is occasionally used to anesthetize the inside of the nose precisely because its vasoconstrictive properties help control bleeding. In any case, you did not get cocaine when you were 7 years old!

A few minutes later, your hand was numb enough to insert the larger needle associated with an intravenous catheter. This numbing effect resulted from blocking voltage-gated ion channels in the nerve cell membrane. In recent years, the importance of ion channels in processes as diverse as taste, nerve signal propagation, and regulation of cellular osmotic pressure has become increasingly well recognized. In fact, Roderick MacKinnon shared the 2003 Nobel Prize in Chemistry for his structural studies of ion channels (see Figure 4.2 for the structure of the potassium ion channel elucidated by MacKinnon and coworkers). In the case of lidocaine and other local anesthetics, the channel in question normally allows passage of sodium cations from outside to inside the nerve cells. This is a critical ion transfer process that normally serves to build up an action potential (charge separation) across the membrane, something that is required for the propagation of nerve impulses first at the local level and then ultimately on to the brain.

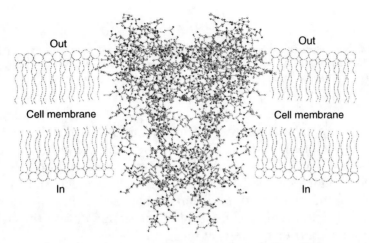

**Figure 4.2.** *Structure of the potassium ion channel as elucidated by MacKinnon and coworkers, who received the Nobel Prize for their efforts. It has been modified to show the approximate placement of the channel within the membrane. Note the two bound potassium cations, whose motion through a functioning channel is thought to lead to ion flow. Such ion flow is essential for a number of biological processes, including the nerve signal propagation impeded by topical anesthetics such as lidocaine and cocaine. This figure was kindly produced by Mr. Dustin Gross, a doctoral student working with Jon, using the published coordinates.*

In simplest terms, lidocaine blocks nerve transmission by acting as a sodium cation surrogate. It is only active in its protonated ammonium form. This is because the neutral or freebase form lacks the positive charge that is an inherent feature of the sodium cation. Fortunately, lidocaine, like most so-called tertiary amines, is basic enough to be protonated at physiological pH (generally 7.4; more on this later). (This acid–base reaction is shown in generalized form in Equation 4.1, where "R" represents a generic substituent and the ":" represents an electron lone pair.) Protonated lidocaine also has the right complement of size and shape to block the sodium channel and is sufficiently hydrophobic that it is taken up in nerve cells.

$$R_3N: + H_2O \rightleftarrows R_3NH^+ + OH^-. \qquad (4.1)$$

To maintain solubility, and because it provides the active protonated form, lidocaine is administered as the hydrochloride salt. Interestingly, when cocaine is used as a local anesthetic, it too is administered as the hydrochloride salt. As with lidocaine, it is this form that produces local numbing and prevents sensations of pain. Since it is a salt, this form of cocaine is highly water soluble. These

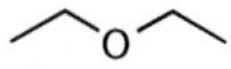

**Figure 4.3.** *Structure of diethyl ether, the first potent anesthesic. Because ether is so lipid soluble, induction and recovery from ether anesthesia both take a long time. Ether is also flammable, and it was responsible for explosions that killed many surgical patients before nonflammable anesthetics were developed.*

kinds of compounds, referred to by chemists as hydrophilic (loving water) or lipophobic (hating lipids),[1] do not easily cross the so-called blood–brain barrier. The blood–brain barrier consists of a layer of lipid-containing cells with tight junctions that protects the brain by preventing the rapid entry of water-soluble species. In contrast, small lipophilic (organic soluble) molecules can readily diffuse through the hydrophobic barrier created by this set of cells.

The freebase form of cocaine is an example of a lipophilic (organic soluble) small molecule. It thus passes through the blood–brain barrier and elicits its psychoactive effects far faster than the corresponding hydrochloride salt. This form of cocaine is therefore prized among drug addicts. In fact, there are famous examples of individuals who have tried to make this form of cocaine with disastrous consequences. This is because they used diethyl ether (generally referred to as "ether"; see Figure 4.3 for structure), a highly flammable substance, to isolate the freebase form after deprotonation of the more readily available cocaine hydrochloride. It is an intriguing coincidence that ether, which is used to purify the first local anesthetic, cocaine, was also the first potent inhalational anesthetic. We will discuss inhaled anesthetics further when we get to that part of your anesthetics. But first, we need to consider a bit more closely the structure of lidocaine and cocaine.

Lidocaine hydrochloride and cocaine hydrochloride are both lipophilic (soluble in lipids) and water soluble. This seeming contradiction in solubility characteristics arises from the fact that the chloride anion is not tightly bound in water; instead, at physiological pH, these compounds exist in the form of large, "greasy" cations. This is an important feature that you can easily tell by looking at their respective chemical structures (see Figure 4.1). What you may not appreciate, though, is that the protonated forms of lidocaine and cocaine are

---

[1] Hydrophobicity is a term for compounds that are organic soluble rather than water soluble. It is often used interchangeably with lipophilicity, that is, soluble in lipids. Lipophobicity, which is generally synonymous with hydrophilicity, denotes compounds that are water soluble.

effective as local anesthetics because they strike an appropriate balance between hydrophobicity and hydrophilicity.

The balance between hydrophobicity and lipophobicity is a recurring theme in drug development. It is particularly significant in the development and use of anesthetics. In the case of local anesthetics, increasing hydrophilicity speeds the onset of the pain-preventing "block" because the drug moves rapidly through the extracelluar fluid, which is mostly water. However, the duration of action is relatively short because the drug binds poorly to lipophilic nerve tissues. In your case, Dan chose the relatively hydrophilic drug lidocaine (as its hydrochloride salt) because it has a fast onset and short duration, which is what you needed.

The balance between hydrophobicity and hydrophilicity likewise helps explain the differences between several well-known over-the-counter nonsteroidal anti-inflammatory drugs. For example, ibuprofen, which is rapidly metabolized to relatively hydrophilic species, is fast acting, whereas naproxen is more lipophilic and thus has a slower onset but longer duration of action (see Figure 4.4 for structures).

The balance between hydrophobicity and hydrophilicity also determines the clinical characteristics of opioid analgesics, compounds that reduce pain without reducing consciousness (much). Dozens of opioids are in routine clinical use, and one reason so many are available is that they have different onset times and durations of action. This allows clinicians to choose the right one for a given circumstance. Most opioid drugs interact with a single type of receptor, the μ-opiate receptor, which, like most cellular components, is lipophilic. Consequently, the primary determinant of onset time is lipophilicity, just as it is for local anesthetics and nonsteroidal anti-inflammatory drugs.

For example, hydromorphone (brand name Dilaudid) is hydrophilic (like lidocaine and cocaine, it has a readily protonatable nitrogen and, in fact, is generally administered as the hydrochloride salt). This analgesic, the structure of which is shown in Figure 4.5, has a

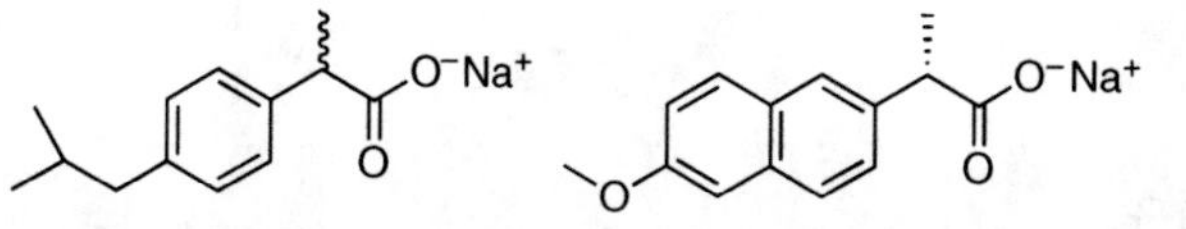

**Figure 4.4.** Structures of ibuprofen and naproxen. Ibuprofen is rapidly metabolized to relatively hydrophilic species and is fast acting. In contrast, naproxen, which contains a hydrophobic naphthalene subunit, is more lipophilic; it is thus slower acting with a longer duration of action.

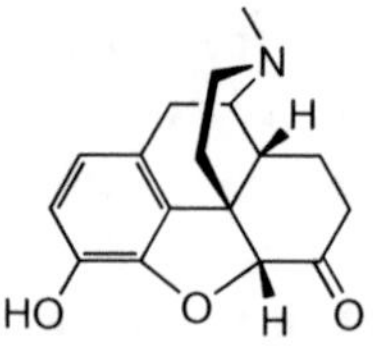

**Figure 4.5.** Structure of hydromorphone, an opioid narcotic. Nearly all opioids bind the μ-opioid receptor. There is no physiological or clinical difference between legal opioids, such as morphine, and illegal ones, such as heroin. In fact, heroin was thought to be a nonaddictive analogue of morphine when it was first developed.

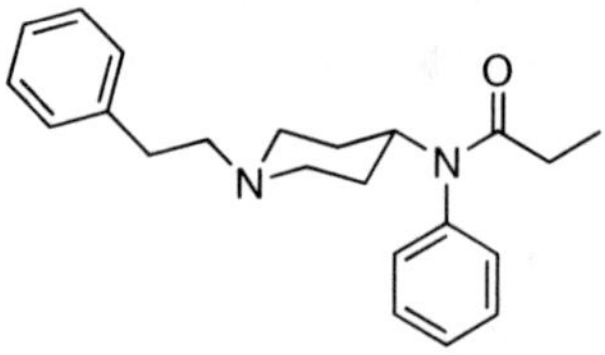

**Figure 4.6.** Structure of the lipid-soluble synthetic opioid analogue fentanyl.

slow onset (about 20 minutes to peak effect even after intravenous administration); however, it then lasts many hours. The reason onset is delayed is that this relatively water-soluble drug can't easily cross the blood–brain barrier. However, once it gets into the brain, the drug remains available to interact with μ-opioid receptors, giving it a prolonged action. In contrast, the lipid-soluble synthetic opioid analogue fentanyl (Figure 4.6) has an onset time of about a minute because it rapidly passes through the blood–brain barrier. However, its analgesic properties last only 10–15 minutes because the drug is so soluble that it redistributes (i.e., is "soaked up") by fat tissues that don't have opioid receptors. This nonspecific absorption terminates its physiological effect. Otherwise, pain relief would last for hours, since the drug is only slowly metabolized. Thus, while the duration of action of opioids is determined by relative water–lipid solubility, some additional mechanisms come into play. Obviously, the anesthesiologist needs to appreciate this or the patients would suffer some very unpleasant consequences!

## MONITORS

During the few minutes required for your hand to become numb, Dan outfitted the middle finger on your right hand with a pulse

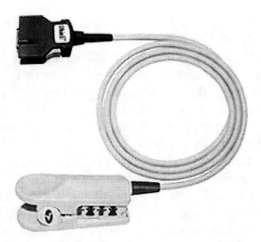

**Figure 4.7.** A pulse oximeter probe, a two-wavelength spectrophotometer that measures the saturation of oxihemoglobin in blood. It is probably the single most important anesthesia monitor.

oximeter (see Figure 4.7). This device is, in essence, a two-wavelength spectrophotometer and uses differential absorption in the red portion of the visible spectrum (660 nm) and in the near infrared (IR) (typically 940 nm) to monitor the relative concentration of oxy- and deoxyhemoglobin in arterial blood. You may recall that arterial blood is a deeper red than venous blood; this is because arterial blood is oxygenated and absorbs light at 660 nm less effectively than the deoxy form (the converse is true at 940 nm). In principle, the Beer–Lambert law you learned last year in freshman chemistry (namely, that the optical intensity of an ideal solution is a linear function of concentration) could be used to calculate the absolute concentration of either oxy- or deoxyhemoglobin from the amount of light passing through your finger at a single wavelength. However, the presence of other absorbing species (including nail polish!) renders such measurements inaccurate. Fortunately, by using two wavelengths, where the absorbance of the oxy and deoxy forms of hemoglobin differ substantially (in the visible and near IR as noted above), it is possible to calculate the ratio of these two species with relatively little interference. This allows the extent of blood oxygenation to be monitored in real time. The pulse oximeter is probably the single most important of the dozens of monitors anesthesiologists use routinely.

Typically, hemoglobin (the oxygen carrier in arterial blood) is nearly 100% saturated when healthy people breathe air under normal conditions of temperature and pressure (i.e., near sea level). Reduced saturation indicates that something is seriously wrong. Unfortunately, a pulse oximeter is not a fail-safe measure of respiratory function. For instance, a patient can have an excessive partial pressure of carbon dioxide in arterial blood and still have 100% oxygen saturation, especially when receiving supplemental oxygen. A pulse oximeter also fails to provide an alert for CO poisoning.[2] This is because the CO- and $O_2$-bound forms of hemoglobin are characterized by a similar bright red color. Needless to say, these are things anesthesiologists have to keep in mind!

Given the limitations of pulse oximetry, you were subsequently attached to an expired gas monitor. (Here, expired refers to breath exiting your lungs and doesn't mean to imply that the monitor was past its "use by" date!) This monitor uses only a single wavelength of IR light, corresponding to a carbon–oxygen vibration to assess directly the amount of $CO_2$ in your breath. While technically simpler, the principle is the same as in the pulse oximeter, namely, the greater the concentration of carbon dioxide, the stronger the absorption (IR light in this case; $CO_2$ has an intense asymmetric stretch—a motion that distorts the resting form from an O-C-O arrangement to an O-C—O structure—at 2345 cm$^{-1}$).[3] The use of calibration curves then allows for an assessment of $CO_2$ partial pressures.

Expired carbon dioxide partial pressure is important because it measures the adequacy of ventilation and provides a measure of blood pH; this is because expired gas from the lungs is in equilibrium with blood passing through the lungs. (Note that blood serves to carry $CO_2$, produced from aerobic metabolism, to the lungs in the form of carbonic acid, $H_2CO_3$.) Since deviations of only a few tenths of a pH unit in blood pH can be either the cause or the effect of serious disease, you can imagine how this is useful information to

---

[2] Victims of smoke inhalation often have high CO concentrations, which can be lethal since CO almost irreversibly binds hemoglobin, preventing oxygen from binding. However, CO is present in the breath of healthy patients since it is formed in small quantities from the natural breakdown (catabolism) of heme, the iron porphyrin prosthetic group in hemoglobin. During anesthesia, CO can be formed from the unwanted breakdown of certain inhalational anesthetics in the $CO_2$ trap discussed in the text proper. Although potentially serious, this latter complication can be avoided through the appropriate choice of anesthetic, flow rates, and, most importantly, not letting the $CO_2$ absorber go dry.

[3] Interestingly, the corresponding symmetric O-C-O stretch is weak. This motion, which converts the resting form from O-C-O arrangement to an O—C—O structure, does not lead to a change in the permanent dipole mode and is thus forbidden. You will learn the basis for this rule of thumb next year when you study spectroscopic selection rules in your physical chemistry course.

have in the operating room—especially as it can be monitored in real time and noninvasively. In fact, isolated changes in expired $CO_2$ partial pressure can be a first warning sign that something is going wrong under anesthesia. As detailed below, this proved true for you!

In performing anesthesia, it is important not just to monitor the levels of $CO_2$ but also to remove it. This is because exhaled air from the patient, containing the administered anesthetic, carbon dioxide, and so on, is recycled (with oxygen added as necessary) through a closed or semiclosed circuit. This is mostly done for reasons of cost. While a few inhaled anesthetics, such as nitrous oxide (discussed below), are inexpensive, most potent inhalational anesthetics are pricey. They are also "greenhouse" gases. And finally, operating room personnel aren't especially interested in breathing low concentrations of waste anesthetics all day at work.

The setup used to recycle inhaled anesthetics in the operating room is called a "circle system" (see Figure 4.8). It recycles exhaled gases and feeds them back to the patient after removing the carbon dioxide and adding enough oxygen to compensate for the amount used by the body's metabolism. Two one-way valves keep flow unidirectional within the circuit.

A critical part of the circle system is the $CO_2$ scrubber (Figure 4.9). Since high partial pressures of $CO_2$ make blood dangerously acidic, it is important to remove it from the circulating air stream. Carbon dioxide is usually removed from the circuit by passing exhaled gas through a container of aqueous alkali or alkaline earth base (typically NaOH, KOH, $Ca(OH)_2$, $Ba(OH)_2$, or mixtures thereof); this

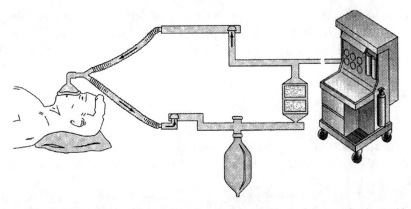

**Figure 4.8.** *A representation of the circle system that is used in modern anesthesia machines. Its cardinal feature is that exhaled gas is recirculated to the patient after oxygen is added and carbon dioxide is scrubbed by soda lime or a similar absorbent. Reproduced with permission from Morgan et al. (2005).*

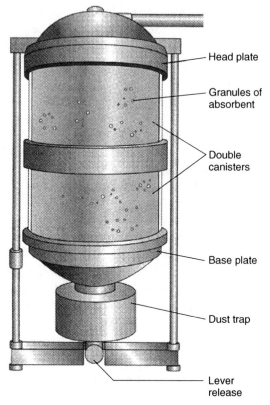

Head plate

Granules of
absorbent

Double
canisters

Base plate

Dust trap

Lever
release

**Figure 4.9.** *Picture of a $CO_2$ scrubber used in anesthesia. Reproduced with permission from Morgan et al. (2005).*

serves to trap the $CO_2$, first in the form of carbonic acid and then in the form of carbonate anion (a species that is, of course, not volatile), as shown in Equations 4.2 and 4.3 for the specific case of $Ca(OH)_2$:

$$CO_2 + H_2O \rightleftarrows H_2CO_3, \tag{4.2}$$

$$H_2CO_3 + Ca(OH)_2 \rightleftarrows CaCO_3 + 2H_2O. \tag{4.3}$$

Anesthesiologists appreciate that the simple acid–base reactions are highly exothermic. Consequently, a properly functioning system is warm to the touch.

A similar $CO_2$-scrubbing system is used in submarines because they would quickly run out of oxygen if it were necessary to discard air just to get rid of carbon dioxide. Furthermore, continuously releasing large amounts of gas could reveal the sub's position! In submarines and during anesthesia, it is important to make sure that the capacity of the $CO_2$ scrubbers is not exceeded. To help guard against this life-threatening occurrence, an acid–base indicator is added to the bed, which "kicks in" and provides a visual alert when the capacity of the trap is nearly exceeded. (This works because, as noted in the equation, $CO_2$ is the anhydrous form of carbonic acid; so, as the scrubber capacity is exceeded, the pH drops.)

## GASES

After your hand was numb and the intravenous catheter was inserted, Dan injected the barbiturate sodium thiopental (brand name Pentathol). Thiopental (see Figure 4.10 for structure) is a fast-onset, short-acting drug that induces coma by reducing nerve-to-nerve transmission at synapses in the brain. At synapses, chemical messengers (neurotransmitters) are released by one nerve and carry a signal to the next nerve across a tiny gap. The neurotransmitters activate receptors on the receiving nerve. Neurotransmission is far from fully understood—and this is an area where chemistry contributes significantly to ongoing research efforts. Based on recent studies involving other intravenous anesthetics, it seems likely that thiopental works by binding to the gamma-aminobutyric acid (GABA) subtype A receptor. This receptor is an ion channel, and when its ligand, GABA, binds, the channel temporarily opens and admits anions into the nerve cell. This creates a negative potential that prevents the cell from producing an excitatory electrical pulse. Intravenous anesthetics thus prolong channel opening, which prevents the nerve from transmitting signals necessary to maintain consciousness. A schematic representation of a synaptic junction is shown in Figure 4.11.

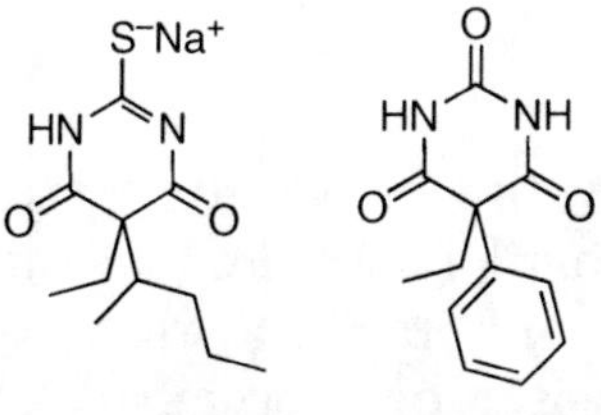

**Figure 4.10.** Structure of sodium thiopental (left). This compound is a derivative of barbituric acid, a compound first prepared by von Baeyer as shown in Scheme 4.1. While barbituric acid itself is devoid of physiological effects, many derivatives, such as phenobarbital (right), have an important history as drugs, as detailed in the text.

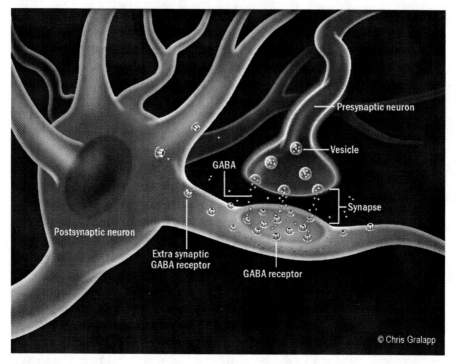

**Figure 4.11.** *Diagram of a synaptic junction. Nerve cells communicate with one another by releasing tiny packets of messenger chemicals, such as gamma-aminobutyric acid (GABA). These "neurotransmitters" diffuse across the tiny gap and activate receptors on the adjacent nerve. This figure was created by Chris Gralapp and is reproduced with permission.*

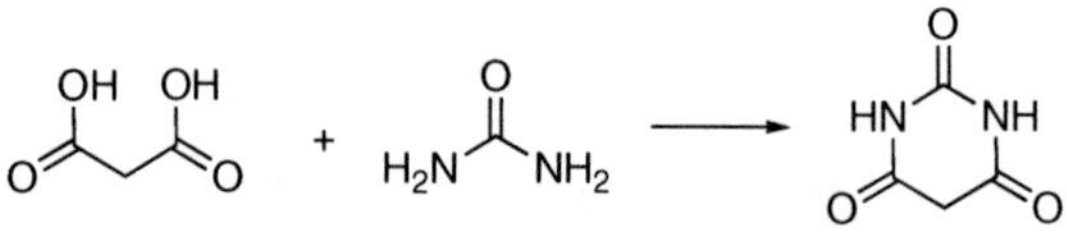

**Scheme 4.1.** *Synthesis of barbituric acid as first carried out by von Baeyer. Subsequently, it was discovered that the diester form of malonic acid (the compound on the left) and its derivatives could likewise be condensed with urea (the compound in the middle), often in better yield.*

Thiopental is a derivative of barbituric acid, a large class of compounds that contain a pyrimidine core (as found in nucleic acids, such as cytosine, thymidine, and uracil). It was discovered by the German chemist Adolf von Baeyer on December 4, 1864—the feast of St. Barbara and hence the name of the drug—and was prepared by combining urea with malonic acid (see Scheme 4.1). While barbituric acid itself is not physiologically active, many of its derivatives (which are generally easy to make) are. Phenobarbital, for instance, has sedative and hypnotic properties. However, it has largely been superseded by safer benzodiazepine-type drugs (e.g., Valium). Meanwhile, sodium thiopental, the drug you received, has been used in low doses as a so-called truth serum. The theory is that in just the right dose, thiopental will disinhibit parts of the brain that provide high-level control while retaining consciousness and memory. Although popularized in spy movies, there is no evidence whatsoever that it works better than a couple of martinis!

Dan administered thiopental knowing that he would soon follow up with an inhaled anesthetic. This switch was made for a very important reason: thiopental is short acting and you would awaken in 10–15 minutes (while your operation was still in progress) if that was all he gave. Of course he could have given more thiopental, but when given repeatedly, the drug switches from being short acting to being extremely long acting (i.e., days!). As with the opioid fentanyl mentioned above, the switch from short to long duration occurs because in small doses, the drug's action is terminated by redistribution to fat tissue, whereas large doses need to be metabolized, which is a slow process. Switching to another intravenous drug was an option, but the most common approach is to use an inhaled gas to maintain anesthesia. Most inhaled anesthetics, including halothane, which Dan used in your case, are inert fluorinated or chlorinated hydrocarbons (see Figure 4.12). However, xenon (a "noble gas" that undergoes no chemical reactions in your body) is also a perfectly good anesthetic. Inhaled anesthetics differ from other drugs in not being metabolized; instead, they enter and exit the body unchanged through the lungs.

Given that halothane was to be used to maintain your anesthesia, you might ask why Dan didn't just start you off with this anesthetic. Why did he even bother putting in an intravenous catheter and giving thiopental? One answer is that most inhaled anesthetics, including halothane, smell terrible. They are thus unpleasant for patients who haven't already been made unconscious by the administration of an intravenous anesthetic. A second issue is that it takes as much as 15

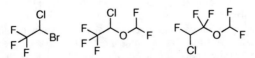

*Figure 4.12.* Structures of three potent inhaled anesthetics: halothane, isoflurane, and enflurane (left to right). Like most inert fluorinated or chlorinated hydrocarbons, inhaled anesthetics are not metabolized; instead, they enter and exit the body unchanged through the lungs.

minutes to induce a sufficiently deep coma with an inhaled anesthetic alone; this is too long to wait in modern operating rooms, which operate at a fast pace. The standard approach is thus to start with an injected anesthetic and subsequently to switch to an inhalational drug.

Now you might wonder how inhalational anesthetics, all physiologically inert gases, make people unconscious. Good question, and the answer is that no one knows! That's right; more than 150 years after the discovery of general anesthesia, how inhalational anesthetics work remains a mystery. However, anesthetic potency is a linear function of solubility in olive oil, a correlation that is maintained over nearly seven orders of magnitude, as illustrated in Figure 4.13. This doesn't mean that the active part of the brain is made of olive oil; instead, oil is just a convenient mimic of the lipid bilayer that surrounds every cell. Most probably, anesthetic gases dissolve in the lipid bilayer and distort the structure, and therefore function, of

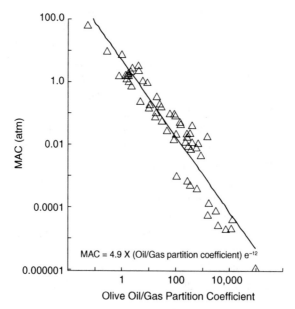

**Figure 4.13.** *Anesthetic potency is a linear function of solubility in olive oil, a correlation that is maintained over nearly seven orders of magnitude. MAC is "minimum alveolar concentration," the partial pressure of a volatile anesthetic that prevents movement in response to surgical skin incision in half of patients. It is the standard measure of anesthetic potency; a MAC of 1.0 defines the average amount required to anesthetize a person. That potency, independent of structure, should be a function of oil solubility over such a wide range is one of the most remarkable findings in all of medicine. But it doesn't mean that the brain is made of olive oil! This figure was generously provided by Edmond Eger II and Dimitry Shnayderman, both of the University of California at San Francisco.*

various cell surface receptors. Confirming this mode of action, or discovering an alternative, thus represents a great challenge for someone of your generation.

Oil solubility doesn't just affect potency; as with so many other drugs, lipophilicity determines the onset and duration of inhaled anesthetics. The more oil soluble the gas, the longer the onset time and the longer the duration. The reason is the lipid-soluble anesthetics are better absorbed into fat. The gases don't produce any physiological effect in fat, but fat serves as a reservoir for the drugs, which then equilibrate with the rest of the body after anesthesia administration is discontinued, preventing the concentration in the brain from decreasing as quickly as it otherwise would. The amount of anesthetic remaining in the body is, unsurprisingly, greater with anesthetics that are lipid (i.e., fat) soluble; hence, their duration is longer.

Slow onset and prolonged action are both disadvantages since anesthesiologists want to be able to titrate drug concentration as needed throughout surgery and want patients to recover rapidly when surgery is done. Paradoxically, recovery times are similar with soluble and insoluble anesthetics if the time in anesthesia is short; in contrast, recovery times after long anesthetic treatments are much shorter when relatively insoluble anesthetics are used. The reason is that during short operations, not much anesthetic—whether lipid soluble or not—has time to be absorbed by fat. In contrast, the fat reservoir becomes increasingly clinically important after several hours. Therefore, it is important to use *short-acting* (less lipid-soluble) inhaled anesthetics for *longer* operations, as illustrated in Figure 4.14.

Nitrous oxide, $N_2O$, was the first inhaled anesthetic and is by far the least oil soluble. ($N_2O$ should not be confused with $NO_2$, which is one of the highly toxic components in smog!) Unfortunately, nitrous oxide isn't strong enough to provide complete anesthesia. It nonetheless remains a popular adjuvant to other anesthetic drugs and has been used for more than a billion anesthesias; it is also used widely in dentistry. The first potent anesthetic was diethyl ether, followed soon thereafter by chloroform, $CHCl_3$. (Interestingly, many in those days thought that people with pain deserved to suffer and that it was immoral to treat pain.[4] Queen Elizabeth was given chloroform to

---

[4] "I think anesthesia is of the devil, and I cannot give my sanction to any satanic influence which deprives a man of the capacity to recognize the law! I wish there were no such thing as anesthesia! I do not think that men should be prevented from passing through what God intended them to endure" (William Atkinson, first president of the American Dental Association).

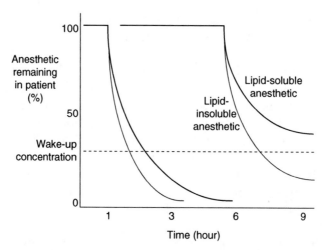

**Figure 4.14.** *Recovery times are similar with soluble and insoluble anesthetics if the times in anesthesia are short; in contrast, recovery times after long anesthetic treatments are much shorter when relatively insoluble anesthetics are used. The reason is that during short operations, not much anesthetic—whether lipid soluble or not—has time to be absorbed by fat. In contrast, the fat reservoir becomes increasingly clinically important after several hours. Therefore, it is important to use short-acting (less lipid-soluble) inhaled anesthetics for longer operations.*

facilitate childbirth, which helped popularize anesthesia.) These were the only completely general anesthetics available for more than 50 years, but both had limitations.

As an anesthetic, ether had two serious limitations. One was that it is so soluble that induction of surgical anesthesia took at least half an hour, and recovery took many hours, depending on the duration of surgery. Of course, ether was still a lot better than the alternative, which was no anesthesia! But a more serious limitation was that ether is highly flammable and its flammability was enhanced by the fact that it was routinely mixed with oxygen since anesthetized patients tend to breathe poorly. The tiniest spark could ignite the ether; the resulting explosion was invariably fatal for the patient, and often for others in the room. As a consequence, even today, there are signs on all operating rooms specifying that flammable anesthetics are forbidden—even though none has been used for at least six decades! Similarly, the vertical drape separating the surgical side of the patient from the anesthetic side is still called the "ether screen" because it was originally designed to keep the ether/oxygen mixture away from the surgeons who could cause sparks with their instrument. (Anesthesiologists, especially when they are trying to annoy surgeons, refer to the ether screen as the "blood–brain barrier"!)

Chloroform was fast acting and nonflammable, but also had two serious limitations. The first was that it caused liver injury in many patients. The second was that a small fraction of patients given chloroform developed lethal disturbances of the electrical conduction system in the heart, leading to sudden death. Fortunately, ether and chloroform have long since been replaced by safer drugs. Three that were popular at the time of your surgery were halothane (the one you received), enflurane, and isoflurane (see Figure 4.12 for structures). None is nearly as lipid soluble as ether, but halothane is nonetheless nearly twice as soluble as isoflurane; enflurane has an intermediate solubility.

The safety of drugs is quantified by their *therapeutic index*, which is the ratio of the toxic dose to the therapeutic dose. For drugs like antibiotics, the ratio is in the thousands, which means that they are extremely safe. This means that you can take more than 100 times the recommended dose of most antibiotics without any harmful effect. Cancer drugs, on the other hand, typically have therapeutic ratios on the order of 3. Cancer drugs are notorious for their life-threatening side effects and for making patients feel extremely ill. In this context, it is worth noting that the therapeutic index for potent inhaled anesthetics is even lower: only about 1.5! In addition, the therapeutic dose varies markedly among individuals and even varies in a given patient over time. This makes inhaled anesthetics the most dangerous drugs in current clinical practice. Their use is thus restricted to highly trained anesthesiologists who remain with patients throughout every minute of anesthesia so they can constantly adjust the dose to match the state of the patient and the progression of surgery.

Appendectomies rarely take more than an hour. Knowing this, Dan chose to maintain your anesthesia with halothane as noted above. (Although extremely rare, halothane has been known to cause lethal liver disease in adults. Fortunately, although for unknown reasons, it never does in children.) Halothane, which, like all potent anesthetics, comes as a liquid, was delivered through a "copper kettle," which is a system that controls vaporization so that the administered dose can be precisely adjusted. Figure 4.15 shows a schematic of a copper kettle; the original copper kettle is shown with an early anesthesia machine in Figure 4.16. Modern anesthesia machines, such as the one shown in Figure 4.17, are far more sophisticated.

These vaporizers are not made of copper to be decorative: vaporization is an endothermic process, which cools the liquid and thus reduces further vaporization. Copper conducts heat extremely well and is thus able to equilibrate its temperature with the ambient

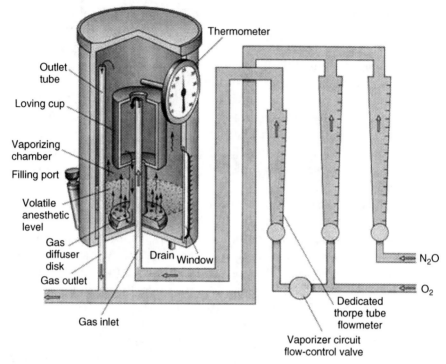

*Figure 4.15. Schematic of a copper kettle, the device that was used to titrate halothane during your anesthesia. Reproduced with permission from Morgan et al. (2005).*

*Figure 4.16. Picture of the first copper kettle, attached to an early anesthesia machine.*

A

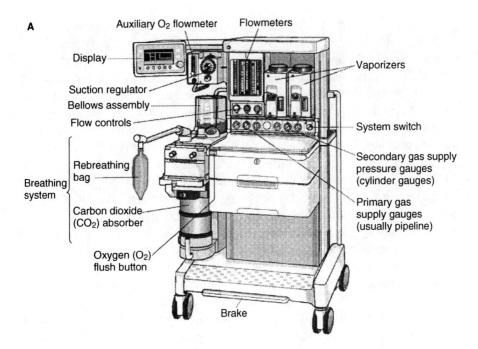

Auxiliary O₂ flowmeter

Flowmeters

Display

Vaporizers

Suction regulator

Bellows assembly

Flow controls

System switch

Breathing system {

Rebreathing bag

Secondary gas supply pressure gauges (cylinder gauges)

Carbon dioxide (CO₂) absorber

Primary gas supply gauges (usually pipeline)

Oxygen (O₂) flush button

Brake

B

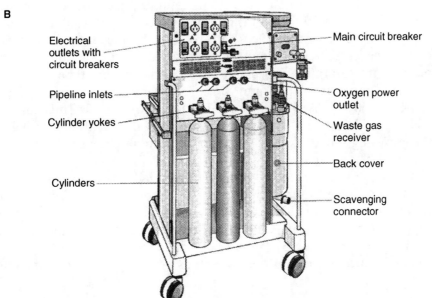

Electrical outlets with circuit breakers

Main circuit breaker

Pipeline inlets

Oxygen power outlet

Cylinder yokes

Waste gas receiver

Back cover

Cylinders

Scavenging connector

*Figure 4.17.* A modern anesthesia machine from Ohmeda (General Electric).

environment; this is important to maintain vaporization at the expected level. The system works by diverting a small amount of carrier gas, such as oxygen, through the kettle where it passes over liquid anesthetic and becomes fully saturated with anesthetic vapor; this small amount of saturated anesthetic (at a concentration that would be instantly lethal) is then diluted with a larger amount of gas to produce the final mixture at the desired concentration. As you might imagine, Dan needed to know his ideal gas laws!

Let's consider the calculations he made. Most patients need an inspired partial pressure of halothane near 1%. The vapor pressure of halothane at a typical operating room temperature of 20°C is 243 mmHg. Also, let's assume that he put 100 mL/min of carrier gas through the copper kettle. That gas would be saturated and thus have a partial pressure of 243 mmHg. UCLA is near sea level, where the atmospheric pressure is typically 760 mmHg. This means that in your case, the concentration of halothane exiting the copper kettle was approximately 32% (i.e., 243/760). To obtain the needed 1% concentration, Dan thus mixed the gas exiting the kettle with approximately 4.55 L/minute of anesthetic-free carrier gas. He calculated this dilution factor by noting that the rate of saturated gas flow exiting the copper kettle was roughly 147 mL/minute. Dan appreciated, as you likely easily can, that the flow rate couldn't be the original 100 mL/minute since "extra" gas molecules were being "picked up" as the result of the saturation process. A bit of simple algebra then gave this increase in volume (per unit time) as being 47 mL/minute since 47/147 × 100 = 32%. To achieve a 1% total flow, this 147 mL/minute had to be diluted with 4.55 L/minute such that the 47 mL/min of halothane exiting the kettle was contained in a total flow of 4.7 L/minute (i.e., 47/4700 × 100 = 1%).

The vapor pressure of isoflurane is essentially the same as that of halothane. However, enflurane has a vapor pressure of 172 mmHg at room temperature. In this case, a 100 mL/minute gas flow through a copper kettle containing liquid enflurane would thus have to be diluted with roughly 3 L/minute of carrier gas to provide a 1% inspired concentration to the patient, as you may want to work out for yourself.

## COMPLICATIONS

After about 20 minutes, Dan noticed that your exhaled carbon dioxide concentration slightly exceeded its normal baseline value of near 40 mmHg; within about 5 minutes, it increased to 45 mmHg, but it then increased to 55 mmHg in just a few more minutes ... and was continuing to climb. Since the body normally tightly controls blood carbon dioxide partial pressure, this increase was more than a little concerning. Dan had to figure out what was going wrong and had to do so quickly. It is not for nothing that anesthesiologists characterize their job, as also do airplane pilots, as consisting of hours of boredom punctuated by moments of panic!

The process by which physicians determine the most likely cause for any particular sign or symptom is called *differential diagnosis*, which is a fancy term for Sutton's law.[5] There are several causes for increased expired carbon dioxide concentration and Dan evaluated each in turn, starting with the most likely. Most often, one of the delicate valves that divert exhaled gases into the carbon dioxide absorber fails. This can happen when moisture makes the diaphragm stick in the open position. However, inspection of the valves (which have clear covers for just this purpose) indicated that both were working normally.

The next most likely cause of an increase in exhaled carbon dioxide partial pressure is exhaustion of the carbon dioxide absorber. This is indicated by a color change, as noted above. Exhaustion also causes the absorber temperature to decrease toward ambient temperature since no further (exothermic) chemical reactions are occurring. The absorber color in your case remained nearly normal. However, the absorber was almost too hot to touch!

That the absorber was so hot was an unusual and alarming sign because it indicated that your body was producing enormous amounts of carbon dioxide, essentially, that you were in an extreme hypermetabolic state. Dan then quickly checked for other signs of hypermetabolism and noted that your heart rate and respiratory rate were both abnormally fast and that your body temperature was rapidly increasing. And finally, he obtained a sample of your arterial blood, which confirmed the classic combination of an excessive carbon dioxide partial pressure and unexpectedly low pH. (Because this "blood gas" is such an important diagnostic test, operating rooms

---

[5] Willy Sutton was a notorious bank robber who eluded capture for decades. When finally apprehended, Sutton was asked why he robbed banks. His famous answer, "That's where the money is," became known as Sutton's law. In the context of medicine, it means that common things are common and that the diagnosis of exotic disease requires a high degree of evidence. Along those lines, medical trainees are told, "When you hear hoofbeats, don't think of zebras."

have machines that can measure blood pH, carbon dioxide, and oxygen concentrations in just a minute or two after completing a blood draw.) Your arterial carbon dioxide partial pressure was 91 mmHg, which is more than twice normal, and your pH was 6.9—a value that is usually lethal—no matter what the cause. The diagnosis was then clear: *malignant hyperthermia.*

Malignant hyperthermia is a rare disease; it is so rare, in fact, that most anesthesiologists see only a single case in a lifetime of practice. It is a genetic disease, triggered by inhaled anesthetics, that causes hypermetabolism in skeletal muscles. When malignant hyperthermia crises are detected quickly and are treated properly, mortality is less than 5%; when diagnosis is delayed or the treatment is inadequate, mortality approaches 100%. You're alive today because Dan recognized this rare syndrome and immediately instituted appropriate treatment.

There are three key treatments for malignant hyperthermia: (1) discontinue the triggering drug (halothane, in your case); (2) increase ventilation to drive excess carbon dioxide from the lungs into the carbon dioxide absorber; and (3) give the drug dantrolene, which reverses the skeletal muscle hypermetabolism. Dan instituted all three treatments. You were critically ill and required intensive care for about 12 hours, but then recovered. Fortunately, malignant hyperthermia is triggered only by inhaled anesthetics; so long as you are not exposed to these gases, the syndrome will have no effect on your life. Since there are other types of anesthesia, you can have surgery perfectly safely should the need arise. But if you do need surgery, don't forget to tell your anesthesiologist that you are susceptible to malignant hyperthermia. And, let's hope she hasn't forgotten her chemistry!

Best wishes,

Jon and Dan

## ACKNOWLEDGMENTS

This chapter is dedicated to our father, Dr. Andrew M. Sessler, PhD, on the occasion of his eightieth birthday. His dedication to a life in science is acknowledged with continued wonder and unceasing gratitude.

The outreach activities in Austin leading to this chapter were supported by the National Science Foundation (grant CHE 0749571).

## FURTHER READING

Morgan, G. E.; Mikhail, M. S.; Murray, M. J.; Larson, C. P. *Clinical Anesthesiology*, 4th Edition, McGraw-Hill Medical, Philadelphia, PA, 2005.

Orser, B. A. Lifting the fog around anesthesia. *Scientific American* 2007, 54–61.

# 5

# *The Green Evolution*

*Carnegie Mellon University*

Terrence J. Collins (Terry), PhD, Hon. FRSNZ, is the Teresa Heinz Professor of Green Chemistry at Carnegie Mellon University (CMU), where he has taught since 1987. Professor Collins is one of the founders of the field of green chemistry. He directs the Institute for Green Science at CMU. He is an honorary professor at and a distinguished alumnus of the University of Auckland in New Zealand. He has received 20 career awards for research and educational leadership. Collins has invented small-molecule peroxidase enzyme mimics, called "TAML® activators," that catalyze the reactions of natural oxidants such as hydrogen peroxide to cleanse water of numerous recalcitrant pollutants and hardy pathogens. He developed the first university course in green chemistry starting in 1992, a course that is now under development to be free and online, aimed at helping to redirect technology toward sustainability. Collins writes and lectures widely about the importance and promise of chemists turning their prodigious inventive talents toward eliminating hazards from chemical products and processes. He is a cofounder of the CMU spin-off company, GreenOx Catalysts, Inc.

*Letters to a Young Chemist*, First Edition. Edited by Abhik Ghosh.
© 2011 John Wiley & Sons, Inc. Published 2011 by John Wiley & Sons, Inc.

Dear Angela,

Thank you for your e-mail. I am delighted that you are intrigued about green chemistry. The world really needs smart young people to become chemists, especially to concentrate on developing sustainable products and processes. In fact, every time a young chemist decides to focus her or his career on dealing with the scientific challenges of sustainability, I think we take an important step toward building the technological dimension of a sustainable civilization. Perhaps you will take such a path and become a green chemist. If so, I hope to meet you one day and welcome you personally as a contributor to green chemistry's vital mission. The most rewarding aspect of my career is that chemistry has given me the incomparable privilege of working with truly brilliant, genuinely wonderful people who are students and postdocs, research chemists, environmental health scientists, environmental health advocates, and government officials. I wish the same good fortune to you.

## CHEMISTRY, SUSTAINABILITY, AND SATISFACTION

You have asked me whether chemistry gives meaning to my life. Yes, resoundingly yes! Chemistry has given me decades of satisfaction and fulfillment, and I feel very fortunate that my career has taken shape the way it has. It has provided me with a wonderful playground for the interplay of imagination and analytical thought. It keeps me on my toes by constantly testing me with fascinating new knowledge. Beyond the technical content, chemistry is integral to pretty much everything that matters in human existence such that it challenges me to continuously grow in my understanding not only of science but also of life. And most significantly, my path in chemistry has brought me into intimate intellectual contact with the most important technological challenges humanity faces as we work to build a sustainable civilization.

The engagement of chemists in sustainability, in its technical *and* other dimensions, is vital for ensuring a good future for our descendants. In recent years, I have realized how forbidding are some of the toxicity surprises associated with commercial chemicals as well as the cultural issues of how we deal with them. So, toward the end of the letter, I will write about how our understanding of hazardous chemicals is undergoing a transformation. Everyday chemicals that once appeared to be benign clearly are not. In some cases, novel toxicities, particularly endocrine disruption, threaten health and the environment (as well as the sustainability of the chemical enterprise) in daunting ways that we are only beginning to appreciate. Endocrine disrupting compounds (EDCs) can alter cellular development at environmentally relevant concentrations to impair living things, including humans. Since I can't imagine that our high technology civilization can have a good future without a strong chemical enterprise, I've added my voice to those who are arguing that it is time to become systematic and conscientious about reducing and eliminating hazardous chemicals. Hazardous chemicals, especially EDCs, are undermining all the truly good things chemists have contributed to society.

There is another dimension we must confront if we are going to make expeditious progress in reducing and eliminating hazardous substances. Some years ago, I found myself wondering why the chemical enterprise, especially as represented by its trade associations, was ignoring and even denying the relevance of so many warnings about EDCs. As I learned more about these issues, I came to the conclusion that money has been trumping health and the environment in totally unacceptable ways. So, I felt compelled to try to understand how this all works and to ask what is good and what is bad about the modus operandi of the chemical enterprise with respect to advancing sustainability. I have tried to explain what I have learned to others through teaching, writing, public speaking, and other mechanisms, including an educational Web site we are developing to make the links between green chemistry and the broader intellectual landscape of sustainability.

I concluded some years ago that there could be no more important personal legacy for me than to contribute to the development of a healthy technology trajectory aimed at a sustainable future. It has become the broader context into which my daily research program fits. For me, it brings an intellectual wholeness to technical depth. I have had the immense personal satisfaction in research of creating TAML activators, green homogeneous oxidation catalysts that are effective mimics of peroxidase enzymes. But beyond planning and executing the research, in my opinion tenured faculty are uniquely situated to speak plainly about their understanding of sustainability matters. In fact, tenure in America was instituted to protect the free speech of faculty so that they could comment on matters of great importance and urgency without losing their jobs. (Tenure was based on the German antecedent of *Lehrfreiheit*, "the freedom to teach.") I have written a viewpoint in *Environmental Science & Technology* about this as a counterpoint to the idea that scientists should just stick to the science and let others decide what to do with it. The technological dimension of our civilization is wrongly constructed in innumerable ways, making everything else unsustainable. It's important that scientists, and chemists in particular, study and commentate on our unsustainable technology trajectories. The more open we are about these challenges, the more likely we will develop sustainable alternatives. The problematic areas include energy, feedstocks, and toxic chemicals. Green chemistry is a discipline where we have to look candidly at our problems if we are going to begin to be able to address them systematically, especially those like endocrine disruption, which count the most. In an authentic green chemistry, scholars will have the daily satisfaction of knowing that what they are doing is truly important to the welfare of mankind and, indeed, to the welfare of all living things.

I got to where I am in an idiosyncratic sort of way. I will explain parts of the journey in the next sections. Since you are beginning your own journey into chemistry, I will focus on my early development and start with some distant memories of what it was like to be a second-year chemistry major in the early 1970s at the University of Auckland in New Zealand. Over the intervening years, I think some key things for undergraduates have changed quite a bit, while others have stayed pretty much the same, and I will reflect on these. I'll briefly discuss the people, ideas, and experiences that attracted me to chemistry and will sketch my PhD and postdoctoral experiences, which comprised the most important foundations of my independent career.

I was drawn into chemistry by the contentment I derived from lab studies. Contentment is the right word for it—working in the undergraduate labs delivered deep satisfaction, and I knew I could happily play in the chemical world for the rest of my life. As you will find in many universities today, in Auckland in the 1970s, we had an introductory lab for first-year students. And that was followed in the second and third year of our 3-year degree by specialized lab courses in organic chemistry, inorganic chemistry, analytical chemistry, radiochemistry, and physical chemistry. I enjoyed all the teaching labs, but especially the inorganic lab, which I started in my second and continued into my third year. The specialized labs were open to us throughout much of the week with a requirement, if I remember correctly, for at least 6 hours. Students could spend as much time as they liked beyond that, and I settled in there whenever I could. I even talked several professors into letting me do an independent holiday project of my own design in the radiochemistry and organic labs. These teaching labs were the place where we got to experiment with real chemicals.

Today, there is rarely as much open access to teaching labs, and I imagine it will be that way for you. But this is balanced by the increased opportunities that undergraduate chemists get to work in research groups. I strongly recommend that you engage in undergraduate research at UCSD because the experience will help you to understand the joys and challenges of research. You will acquire technical skills that will give you a running start when you go on to graduate school. We did not have an undergraduate research program in my day. So instead, it was the fun I had in those undergraduate labs making compounds, growing crystals, and studying properties that led me to want to pursue higher degrees in chemistry.

## LOOKING BACK ON THE EARLY PARTS OF MY OWN CAREER

The two most important people in my scientific development were my PhD and postdoctoral mentors, Warren Roper and Jim Collman. In New Zealand, students wishing to continue on beyond the 3-year bachelor's degree had to start by pursuing a 1-year master's degree complete with a substantial research thesis. Following my bachelor's degree, I was very lucky to be assigned for the MSc degree (yes, we were assigned in those days by the head of department) to the research group of Warren Roper in the field of organometallic chemistry. I quickly become enthralled with making and studying compounds that no one had ever seen before. I studied cyclic carbene complexes for my master's research and was lucky to win a scholarship based on that effort to pursue a PhD with Professor Roper. It was a life-directing experience. My PhD research was focused on thiocarbonyl complexes of ruthenium and osmium. Under Professor Roper's brilliant guidance, I made over 100 new compounds. I identified the first migration reactions of hydride ligands on to the thiocarbonyl ligand, reducing it stepwise all the way to the methylthiolate ligand, which then could be stripped with acid from the metal as methyl thiol (Figure 5.1). Although the mechanism eventually proved

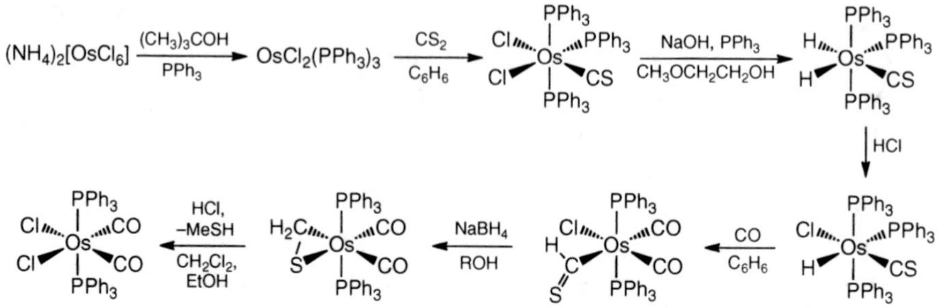

**Figure 5.1.** Stepwise reduction of the thiocarbonyl ligand on osmium.

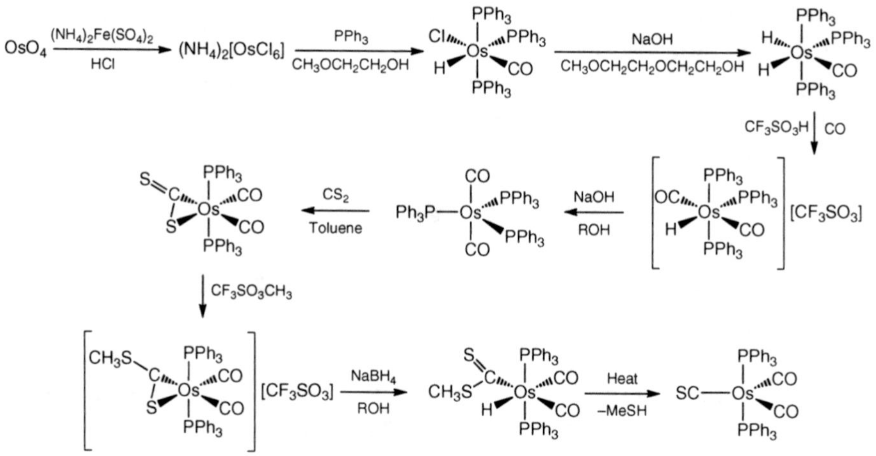

**Figure 5.2.** Synthetic pathway for producing Os(CS)(CO)$_2$(PPh$_3$)$_3$, a zerovalent osmium thiocarbonyl complex intended for use in novel oxidative addition reactions.

to be different, at that time, we considered this thiocarbonyl ligand reduction chemistry to be a model for the sequence of reactions in the catalytic reduction of carbon monoxide to methanol, a process that was of great industrial interest at the time and still is.

But in its execution, my PhD experience was not a smooth ride, at least to start with. I spent the first 18 months pursuing a grueling multistep synthesis of what was to be my key starting material, a zerovalent osmium thiocarbonyl complex (Figure 5.2). The idea was to use this reduced species to study novel oxidative addition reactions. I remember being happy to smell the stink of methyl thiol in the last step as that meant the reaction was working. But I could never get enough starting material to do anything significant with, and the reaction often produced an intractable dark oil. So not much new was coming out of my research efforts. I worked hard to try to turn things around, but I certainly felt like things weren't working most of the time.

Professor Roper is a meticulous scholar gifted with a great memory. He read everything pertinent to his field of interest. One day, he advised a new tack based on something he had just read (first two steps in Figure 5.1) for getting a thiocarbonyl ligand onto osmium. It worked quantitatively, giving me as much thiocarbonyl complex as I needed just three steps away from the starting osmium tetraoxide. From that point on, I could almost do no wrong in the lab, and I was able to complete the reactions of Figure 5.1 along with many others. It's wonderful to experience what research feels like when it is going really well. But I often think the tough first half of my PhD taught me the most significant lesson—*tenacity is critical for success in science*. I graduated in a little more than 3 years—PhD degrees in New Zealand in those days were relatively short compared to contemporary American standards.

The Roper Group in Auckland was a wonderful environment in which to work and study. Very often, the students would eat brown-bag lunches with Professor Roper, and we would all enjoy talking about the issues of the day. Professor Roper was conservative in his outlook on the world. Some of my fellow students and I were more liberal. We would spar happily with Professor Roper on matters of world issues. This led to many enjoyable debates where we argued about the broader significance of things in science and beyond and to playful antics that became the cement of lifelong friendships. These experiences proved to be the underlying foundations of scientific collaborations that proceed to this day between former graduate colleagues now on the faculty (or "the staff" as they call the academics there) in Auckland, especially with James Wright, as well as a great personal friendship with Penny Brothers.

After my PhD was completed in 1978, I was really fortunate to get accepted as a postdoctoral fellow into the research group of Professor James P. Collman at Stanford University. In ways different from Professor Roper's, but just as powerfully, Professor Collman inspired me as a research mentor. Jim Collman has a great sense of the big picture of science and a nose for research areas where world-changing chemistry is waiting to be discovered. In his group, I became engaged in metalloporphyrin projects and learned to become somewhat competent in organic synthesis. I was exposed to the wonderful things you could learn through electrochemical techniques as well as many other research areas that were novel to me. The Collman Group was large and impressively international. It was full of outstanding students and postdocs. The technical skills of the people in the group were diverse and exceptional. I got to learn from people of other cultures about how they look at things in their own unique ways. I was able to experience a large seminar program and to see firsthand and interact personally with many leaders of chemistry of that day. From time to time, I acted as the host for seminar speakers and that was a particular delight. Sometimes, when Professor Collman traveled, I filled in for him and lectured to his large sophomore organic chemistry class. That was a very rewarding challenge. I have to be honest and say that there was quite a bit of culture shock also, but experiencing culture shock is one way to grow quickly.

In the Collman Group, there were many different projects, but most were aimed at modeling oxidizing enzymes. This was a hot area of research at the time, and still is, with many exceptional groups around the world contributing to its development. So, while Professor Collman's students and postdocs were teaching me a growing battery of technical skills, I was also learning from Kevin Smith's brilliant book, *Porphyrins and Metalloporphyrins*, as well as from the papers and reviews of many outstanding chemists in the porphyrin field.

As with my degrees at Auckland University, Stanford was a positive and life-directing experience. My mentors and the people I worked with made all the difference. Professor Collman was so inspiring on the need for creativity in science and the possibilities for doing great things that simply being around him made you believe you could go out of his group and develop new chemistry to advance the world. While I was there, Professor Barry Sharpless (a former Collman postdoc) made the discovery in enantioselective catalytic oxidation chemistry that led to his Nobel Prize, and it was great to talk to him and his group while this was happening. But most of all, Jim Collman knew how to make you believe in yourself as a scientist.

I have tried to incorporate as much as I can of Professor Collman's and Professor Roper's great qualities into the training of my own students and to bring some part of them into the chemistry program at CMU as well. Freedom for self-expression in science is the most treasured ingredient to build into a graduate education. Professor Collman was an avid runner and he would let off steam by running in the beautiful Stanford Hills with any postdocs or students who were game enough to try to keep up with him. And it was in the Stanford Library near the end of my postdoc that I envisioned the research program that has largely occupied my own group for 30 years. I discussed my goals with Professor Collman on our runs, and his enthusiasm for the ideas meant a great deal to me.

## PLANNING AN INDEPENDENT CAREER

Useful homogeneous oxidation catalysts are relatively few in number. Among the most important goals in the porphyrin field in 1980, when I began my independent career at Caltech, was the desire to produce effective small-molecule mimics of the mono-oxygenase and peroxidase enzymes. The Collman Group was concentrating on modeling cytochrome $P_{450}$ enzymes, mono-oxygenase enzymes that activate oxygen to oxidize organic substrates via iron-oxo intermediates. So, in 1980, based on what I had learned under Professor Collman's tutelage, I decided that I would try to develop mimics of the iron peroxidase enzymes that activate hydrogen peroxide. These enzymes have ferric resting states. They react with hydrogen peroxide to produce two oxidizing intermediates, a (porphyrin radical cation)iron(IV)-oxo species two oxidizing equivalents above the resting ferric state called compound I and a (porphyrin)iron(IV)-oxo obtained by a one-electron reduction of compound I called compound II.

I was additionally motivated by thinking that if effective peroxidase mimics could be produced, it might be possible to disinfect water with hydrogen peroxide rather than with chlorine. In the late 1970s, scientists were realizing that chlorine disinfection, fundamentally a complex oxidation process, produces chlorinated by-products. Some of these are carcinogenic—chloroform is of particular concern. The popular press had been alerting people that some cancer cases were likely to be associated with the resulting exposures. So, in 1980, I saw both a "can it be done?" and a human health argument for finding effective, economical, small-molecule peroxidase mimics. The combination was irresistible and sustained me through 15 years of moving closer to my goals, until we did actually achieve functionally useful peroxidase mimics in 1995.

As the years went by, I also learned many other things about toxicity problems associated with the chlorine industry. I have long since arrived at the conclusion that some of chlorine's toxic residuals are so odious that the industry should be fundamentally restructured with a much higher sensitivity being accorded to health and the environment than currently pervades our culture. And especially since we must reduce and eliminate endocrine disrupting chemicals (the chlorine industry is responsible for some key EDCs), many sectors of the chemical industry have major challenges before their eyes and will need a lot of help. But you can read about my ideas and views on these issues if you wish.

## LIGAND DESIGN FOR PEROXIDASE MIMICS

In my independent career, for 6 years at Caltech and for more than two decades at CMU, I set out to develop nonporphyrinic chelating ligand systems to support iron peroxidase-like catalysis. The design questions involved figuring out what properties the ligand systems would need to possess for them to deliver useful catalysts. It seemed reasonable that tetradentate, tetra-anionic chelating ligands with a high overall donor capacity would make high-valent iron-oxo reactive intermediates more accessible such that these could serve as peroxidase-like intermediates in the oxidation of organic substrates. I put the concern of overdoing the stabilization aside and figured we would back out of that if it became a problem. Overstabilization did prove to be a problem with manganese; we made the first stable manganese(V)-oxo complexes, and we did find a way to back out of it, but it was cumbersome. Nothing beats simplicity and as it would prove in our synthetic systems as in the enzymes, iron is special in enabling a certain beautiful simplicity in the broader context of complexity. This is true in iron TAML activator catalytic cycles as it is in enzymatic catalytic cycles, just as a beautiful melody can hold a complex symphony together.

Most importantly, I realized in 1979 that any ligand system and derivative complex that could work for this task would need to be highly resistant to oxidative and hydrolytic decay. We began by designing nonmacrocyclic diamido-diphenolato ligand systems. In the 1970s, Dale Margerum at Purdue University had shown that deprotonated organic amides were strong σ-donors to metals. The key C(O)–N component looked to be oxidatively inert. Oxidative sensitivity looked as if it would reside in the organic groups of the RC(O)–NR′–M system, so amides became the key building blocks for our chelating ligand systems and finding sufficiently oxidatively inert R– and R′–moieties to combine with them became the task. The first 15 years of the project ended up being focused on using iterative design to gain control of the kinetics of the decay processes of the catalyst ligand candidates we were designing.

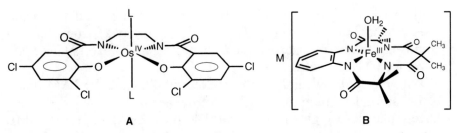

**Figure 5.3.** *(a) The initial ligand system (1980) bound to osmium in our catalyst development program and (b) the prototype TAML activator (1995).*

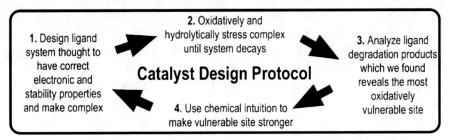

**Figure 5.4.** *The catalyst design protocol that led to TAML activators.*

The first system that my group began developing (Figure 5.3a) was not that different in broad terms from the TAML systems that provided in 1995 a solution to the peroxidase-mimicking challenge (Figure 5.3b). But the two ligand systems are worlds apart in the subtler features of their design. The daydreams in the Stanford Library led to an iterative design protocol for how one might achieve oxidatively robust ligand systems (Figure 5.4), and my group has followed this throughout the intervening 30 years—we continue to use it to improve our catalysts to this day. We quickly learned that iron with the starting acyclic ligands tends to produce dimeric species akin to a chromium dimer that we had characterized by X-ray crystallography. There was an infrared signature in the amide ligand vibrations that marked amide-O coordination when we needed amido-N binding exclusively. But we found that osmium produced hydrolytically stable mononuclear complexes of the N-coordinated type we were looking for, so we first studied ligand oxidative degradation reactions on osmium and coupled this to the iterative design depicted in Figure 5.4. The superior coordinative stability of osmium versus iron assisted us by making ligand hydrolyses inconsequential. By the mid-1990s, we were able to prescribe a set of rules that helped in attaining oxidatively and hydrolytically robust homogeneous oxidation catalysts. At that time, we were one ligand iteration away from the iron complexes that mimic peroxidase enzymes very well. By 1995, we had developed TAML activators.

With molecular weights in the region of 500 Da, ferric TAML activators are indeed miniature replicas of the peroxidase enzymes, which have molecular weights from ca. 40 kDa to ca. 250 kDa. Ferric TAML activators exhibit impressive performance in peroxidase-like chemistry. At miniscule concentrations (nanomolar to low micromolar), ferric TAML catalysts accelerate peroxide chemistry to degrade a wide range of pervasive and recalcitrant chemicals in water. One example is ethinylestradiol, an active ingredient in birth control pills that turns up in environmental waters where it can impair aquatic life at environmentally relevant concentrations. At 83 nM concentration, the prototype ferric TAML catalyst can destroy close to 1000 equivalents of ethinylestradiol (80 μM) within 15 minutes at room temperature with a small excess of hydrogen peroxide over the mineralization requirements. Under these conditions, 1 kg of the ferric TAML catalyst would treat over 20,000 t of water. Another is pentachlorophenol, an extremely persistent pollutant, which is completely destroyed to our limits of detection and nearly mineralized within minutes at room temperature. Importantly, dioxins are not detectable by-products within the limits of the mass spectrometers at the German National Labs for the Environment in Neuherberg in Bavaria—the oxidation of chlorinated phenols often leads to dioxins. The list of degradable compounds includes recalcitrant drugs, pesticides, dyes, aromatic gasoline components, pharmaceuticals, organochlorines, organosulfur compounds, the colored and smelly contaminants associated with pulp and paper mill effluents, and more. Ferric TAML/peroxide also rapidly kills bacterial spores, the hardiest of pathogens, as well as other microbes. In the real world, the idea is to clean waste streams of persistent chemicals and/or pathogens in a green way before they are released to the environment.

Just as my mentors and peers were so important to the early stages of my journey into chemistry, thereafter, my graduate (and undergraduate) students, postdocs, and senior researchers have made daily life in chemistry a great pleasure. It's not that everything is always perfect for any of us. But in the agglomerate, it has been and remains a wonderful journey. I am immensely proud of the achievements of my graduate students and postdocs. Senior members of our team in CMU's Institute for Green Science include Alexander Ryabov (our kinetics guru) and Sushil Khetan (our organic colleague and pesticides, pathogens, and explosives expert). Along the road to TAML activators, we developed much novel coordination chemistry and we collaborated extensively with outstanding scientists, Eckard Münck, Emile Bominaar, James Wright, Dieter Lenoir, Karl-Werner Schramm, and less often with many others.

Our ongoing research remains aimed at designing better oxidation catalysts as we try to improve on the performance of ferric TAML activators. We also work on understanding the mechanisms of ferric TAML activator catalytic reactions, on developing the basic science needed for understanding how they can be used in real-world applications, and on trying to ensure that the catalysts and their processes are truly green. A former Institute research professor, Colin Horwitz, has for some time been focused on developing business opportunities and is now the chief technology officer of GreenOx Catalysts, Inc., the spin-off company commercializing TAML activators. You can read more about TAML activators from the papers we have published, which can be found in the reviews below if you are interested. TAML activators are an important example of green chemistry in action. But for the remainder of this letter, I would like to share with you some thoughts on green chemistry, which I believe will preoccupy your generation of chemists for decades to come.

## GREEN CHEMISTRY

TAML activators are an important example of green chemistry in action. Green chemistry was named and defined by Paul Anastas as "the design of chemical products and processes that reduce or eliminate the use and generation of hazardous substances," and further introduced by Anastas and John Warner in *Green Chemistry: Theory and Practice*. There are several definitions, but the original Anastas definition is my preferred one. Our catalyst design program fitted so well with Dr. Anastas' green chemistry that we became engaged as soon as I heard about it in 1991. Green chemistry, writ large, has become an important part of my life and my group's daily efforts. Today, we are helping to develop the field by working with others on curriculum development and by participating in a unification of green chemistry with environmental health sciences. The nonprofit group Advancing Green Chemistry is leading the construction of this strategic partnership, which has been inspired by Pete Myers, the publisher of *Environmental Health News* and a world leader in bringing the understanding of EDCs to us all. So I will conclude this letter by highlighting what I believe is green chemistry's transformational potential with respect to reducing the adverse environmental and health impacts of chemistry and reorienting it toward sustainability.

We all live in a world where our understanding of the meaning of science and technology for human advancement is changing. In the last several centuries, science and technology have given us new powers humans never had before. In my opinion, Hans Jonas has taught us best how to think about the ethical implications in his brilliant book, *The Imperative of Responsibility: In Search of an Ethics for the Technological Age*. Paraphrasing Jonas, it is easy to see that if you had lived 500 years ago, your own personal power would have been so much less than it is today. In 1500, you would have had to meet people to influence them—today, you can influence people all over the world almost instantly via electronic means. Not long ago, we all witnessed important communications history being made as the people of Iran informed the world of their postelection troubles through thousands of messages and videos sent surreptitiously via innumerable electronic routes. In 1500, you probably would not have ventured far from your home—today, people travel the globe with remarkable speed and comfort. If you lived a long life in 1500, you might have met three or four generations. Most of them would have been residents of the town you lived in, which would probably have been small and protected by robust walls. You would probably have considered the natural world outside the walls to be a relatively dangerous place. You would not have known what pathogens are. Yet they would have been busy culling your town's population. The thought that what you did each day could have had a profound impact on the physical health and welfare of people living 500 years later as well as on the vitality of nature would have seemed odd if not just plain absurd. But today, the idea that what we do each day, individually and collectively, might be impacting the welfare of humans and the ecosphere in 2500 is very realistic. In fact, it is inescapable.

Our civilization has taken many wrong paths in technology development that lead away from a sustainable future. To take a local example, consider the state of Pennsylvania and its posture toward energy. Our state has a population of about 12.5 million people. In October 2008, the Department of Energy ranked Pennsylvania third in carbon emissions (284.0 million metric tons) behind Texas (ca. 25 million people, 625.2 million metric tons) and California (ca. 37 million people, 395.5 million metric tons), so our per capita pollution of the atmosphere is close to as bad as it gets. Pennsylvania has been a big coal state for a long time, but over recent years, a major gas deposit has been found deep underground in a Devonian period sedimentary rock called the Marcellus Shale. Much of this enormous gas field is under Pennsylvania, and it is currently thought that about 10% of this, 50 trillion cubic feet, can be extracted. The "fracking" (fracturing) process employed involves the injection of water, sand, and hundreds of chemicals at high pressure into the shale along horizontal drilling shafts spreading from a central vertical shaft. The shale is fractured along the horizontal tracts first with explosions and then with fracking fluids injected under high pressure. The injected chemicals in these fluids include carcinogens and developmental disruptors. A considerable percentage returns to the surface with the gas and is collected in open pits, leading to contamination of water, air, and soil. Some people are getting sick quickly. Many people can light a match at their kitchen sink and the water coming from the tap will flare. But the amounts of money to be made are large and, for now, the gas mining is mostly seen locally as a boon to the economy. The hard questions associated with climate change, water and air quality, and health are drowned out by the enthusiasm engendered by the short-term economic potential. So, Pennsylvania will likely develop the Marcellus deposits to prop up the local economy. Already, the exploration rights on State lands are nearly all sold. While we will get the energy and the near-term income, low-lying islands and coastal lands throughout the world will likely continue going underwater. Our activities with coal and natural gas will likely contribute to the melting ice and the rising waters. Pennsylvania's fresh water and lands will be polluted by the fracking chemicals and released minerals, some of which are highly toxic, probably for a very long time to come. If history is any guide, the gas miners will not pay for these severe long-term penalties to health and the environment.

In 2005, Congress gave the gas miners a pass on the Clean Water Act and other acts that protect health and the environment. Changing this back is a critical thing to do for health and the environment. The way our culture works, I believe that our best hope of a better course (and leaving the gas right where it is) lies in the ability of our national leadership to keep energy prices down by rapidly promoting sustainable solar energy, where chemists have so much to offer. If the solar trajectory takes root quickly enough, Pennsylvania could contribute by turning some of its vast fossil carbon resources toward expanding manufacturing and producing the equipment the solar transformation requires. Pennsylvanians are proud of the fact that all the parts of large windmills are made in the state—let's hope we will contribute strongly to the solar-to-electric and other renewable energy developments.

Jonas captures our energy and other sustainability challenges by urging that we advance our general mentality. In powerful prose, he calls for the development of a new ethics emphasizing our responsibility to use our greatly expanded powers for the betterment of future humans—"responsibility is a correlate of power." He proposes a new ethics, which today we might call *sustainability ethics*, for handling the powers we wield through science and technology "to save the survival and humanity of man from the excesses of his own power." Its supreme principle is that one should "Act so that the effects of your action are compatible with the permanence of genuine human life." Because chemistry is so much a part of our newfound power and especially because we live in a world where synthetic chemicals are endocrine disruptors, sustainability ethics and chemistry are intimately and inseparably intertwined. So, in my green chemistry class, I ask students to read and evaluate Jonas' *The Imperative of Responsibility* and also Colborn, Dumanoski, and Myers' *Our Stolen Future* and Markowitz and Rosner's *Deceit and Denial*. Each year, I am amazed by the eloquence and insight expressed in the essays I get to read.

Always remember that the core idea of green chemistry is simple. Green chemists design against hazards to build sustainable technologies. To look after the world for future humans, we must develop safe energy, especially by enacting already technically attractive conversions of solar to electric and chemical energy that will only get better with time. We must develop the chemistry of renewable feedstocks to make high-value products from recently dead as opposed to fossilized plant matter. And we must reduce or eliminate hazardous compounds from the technology base. The more serous the hazard is, the more important the green chemistry design challenge is. Obviously, green chemists have to understand the hazards they are designing against at a deep level. And this requirement presents novel and exciting tasks, especially for the more complex toxicities, such as endocrine disruption, that require multidisciplinary expertise. So, green chemists are working with environmental health scientists to build an alliance aimed at bringing science-based understanding to rid the world of developmental disruptors.

Angela, I believe things will go well for our civilization because of young people like you. Your generation will oversee a transformation of chemical research in the three general areas discussed above. And when you become chemistry professors, I believe you will consider it quite natural to teach sustainability ethics in chemistry classrooms. I am sure you will continue to expand toxicity and ecotoxicity education in the core chemistry curriculum to better equip chemists to design with human health and the environment in mind. As green chemistry expands and more young people climb on board, yours will be the most important generation in building the technological dimension of a sustainable civilization.

Best regards,

Terry

## FURTHER READING

Anastas, P. T.; Warner, J. C. *Green Chemistry: Theory and Practice*, Oxford University Press, Oxford, New York, Tokyo, 1998.

Colborn, T.; Dumanoski, D.; Myers, J. P. *Our Stolen Future*, Penguin Group, New York, 1996. Available at http://www.ourstolenfuture.com.

Collins, T. J. Green chemistry. In *Macmillan Encyclopedia of Chemistry*, Vol. 2, Lagowsky, J. J. (ed.), Simon and Schuster Macmillan, New York, 1997, pp. 691–697. Available at http://www.chem.cmu.edu/groups/collins/.

Collins, T. J. Toward sustainable chemistry. *Science* 2000, *291*, 48–49.

Collins, T. J. Persuasive communication about matters of great urgency: Endocrine disruption. *Environmental Science & Technology* 2008, *42*, 7555–7558.

Collins, T. J.; Walter, C. Little green molecules. *Scientific American* 2006, *294*, 82–90.

Collins, T. J.; Khetan, S. K.; Ryabov, A. D. Iron-TAML catalysts in green oxidation processes based on hydrogen peroxide. In *Handbook of Green Chemistry*, Anastas, P. T.; Crabtree, R. H. (eds.), Wiley-VCH Verlag GmbH and KgaA, Weinheim, 2009.

Grossman, E. *Chasing Molecules: Poisonous Products, Human Health and the Promise of Green Chemistry*, Island Press, Washington, DC, 2009.

Jonas, H. *The Imperative of Responsibility: In Search of an Ethics for the Technological Age*, University of Chicago Press, Chicago, 1984.

Markowitz, G.; Rosner, D. *Deceit and Denial: The Deadly Politics of Industrial Pollution*, University of California Press, Berkeley, CA and Los Angeles, 2002.

# Part II

*Chemistry and
the Life Sciences*

# 6

## *Thinking Like an Enzyme*

**Judith P. Klinman**
*University of California, Berkeley*

Judith Klinman received her PhD in chemistry at the University of Pennsylvania in 1966. She was a postdoctoral fellow at the Weizmann Institute for Science, Israel, from 1966 to 1967. In 1968, she returned to Philadelphia and proceeded to spend 10 years at the Institute for Cancer Research in Philadelphia, first as a postdoctoral associate and later as a research staff member. In 1978, she moved to the University of California, Berkeley, where she is now chancellor's professor of chemistry and of molecular and cell biology.

Dear Angela,

A few weeks ago, my sister and I prepared a luncheon for a group of women with whom we had graduated from high school 50 (!) years ago. We had gathered from all corners of the United States—California, Washington, DC, New York, New Jersey, and Pennsylvania—to compare notes and just to enjoy each other. I was the only scientist in the group—a situation I am sure will be different in 2056. One thing we all sensed was our comfort with each other. Gone were the tension and (yes) competition from many years ago—we had become who we were. ...

As you begin your journey in life, the decisions may seem overwhelming. My hope is that you will have a few wonderfully passionate and knowledgeable teachers who provide you with that "eureka" moment in which your future professional path becomes clear. This happened to me first in high school when an outstanding and involved chemistry teacher made me turn away from a major in French literature to chemistry. A second turning point occurred at the University of Pennsylvania, where, as an undergraduate, I had the chance to work as a lab technician at the Johnson Foundation, a world-renowned center for research in biochemistry and biophysics. This experience sparked a lifelong desire to understand biology on the basis of chemical principles. Finally, I was fortunate to obtain a postdoctoral appointment with Irwin Rose (Nobel Laureate in Chemistry in 2004) at the Institute for Cancer Research. It was here that my journey into "how to think like an enzyme" began.

## ENZYMES—CATALYSTS FOR TRANSFORMING MOLECULES ON AMAZINGLY FAST TIMESCALES

Enzymes are everywhere—most notably as the 24/7 workhorses within all living cells. But pick up a box of detergent and you are also likely to find an enzyme that has been added to help break down the grease stains on your favorite white dress. If you wear contact lenses, as I do, the cleaning solution may contain an enzyme that works to hydrolyze deposited protein. Commercial enzymes have made their way into feedstocks to aid farm animals in the digestion of their food. It appears that the potential applications of enzymes are endless, only limited by the resourcefulness of the research scientist and the curiosity and energy of the entrepreneur. As workers in the pharmaceutical industries have demonstrated over many decades, enzymes are often the primary target in drug design.

The most dramatic feature of enzymes is that they are highly selective catalysts, allowing reactions that would normally require years in their absence to take place in milliseconds. The upper limit for the extent that an enzyme can accelerate a reaction is estimated to be $10^{20}$, about the same order of magnitude as the number of observable stars in the universe! The way in which enzymes are able to bring about such enormous rate accelerations has fascinated researchers for over half a century. The goal is to be able to understand the features of enzyme catalysis to such an extent that we are able to design a protein-based catalyst from first principles. Despite much effort, this goal continues to elude researchers.

Let's take a look at an enzyme from two perspectives—first in its entirety and then as a close-up of the active site. Although biological catalytic activity has been found for polymers of nucleic acids, most enzymes are protein polymers of amino acids that link together to form very large molecules with molecular weights in the range of 10,000–100,000 Da. Despite the presence of discrete structures within a protein, the overall structure may look quite haphazard, kind of like a glob, and what you are looking at in Figure 6.1a is in fact called a "globular" protein. This may seem surprising—that a structure with so much important work to do could look so irregular—perhaps leading to the (correct) suspicion that the irregular shape may play a role in function.

In contrast to the expansiveness of the entire protein, the active site region that binds the substrate and catalyzes its conversion to product represents a very small fraction of the total protein (Figure 6.1b). Much of the work during the twentieth century was focused on being able to visualize proteins in three dimensions and, as a consequence, to identify the residues that interact directly with the bound substrate at the active site. These activities led to the recognition of additional roles for small inorganic and organic cofactors that are the major constituent of your "once-a-day" vitamin pill. Such cofactors bind at or near an enzyme active site, acting in concert with numerous protein side chains to effect catalysis. In fact, as a result of such studies, we can now write out a reaction scheme that describes the pathway for the making and breaking of bonds within a large number of enzyme classes. However, note that this level of understanding is quite different from our ability to understand and reproduce the precise origins of the catalytic rate acceleration.

One feature that has become very obvious from the inspection of an enzyme structure in the presence of a substrate (or a substrate analogue that is prevented from reacting) is the enormous number of interactions that are possible between the protein and the bound ligand (Figure 6.2). In some instances, X-ray structures reveal the presence of up to 20 discrete and highly directional interactions that may include hydrogen bonds, salt bridges, pi-stacking interactions, and so on. In this context, let us consider a hypothesis for how

A

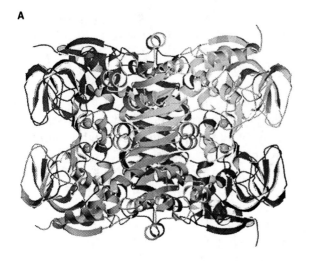

B

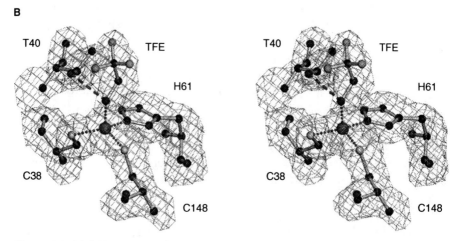

***Figure 6.1.*** *(a) Ribbon diagram for a thermophilic alcohol dehydrogenase. This illustrates the arrangement of the four subunits and the overall size of the molecule in relation to bound zinc atoms, shown as solid spheres. (b) Catalytic zinc site of the thermophilic alcohol dehydrogenase. The catalytic Zn atom is in a distorted tetrahedral coordination by side-chain heteroatoms of residues Cys[38], His[61], and Cys[148]. Trifluoroethanol (TFE) is a stable (unreacting) analogue of the enzyme's substrate. Source: Ceccarelli, C.; Liang, Z.-X.; Stickler, M.; Prehna, G.; Goldstein, B.M.; Klinman, J.P.; Bahnson, B.J. Crystal structure and amide H/D exchange of binary complexes of alcohol dehydrogenase from* B. stearothermophilus: *Insights to thermostability and cofactor binding.* Biochemistry *2004, 43, 5266.*

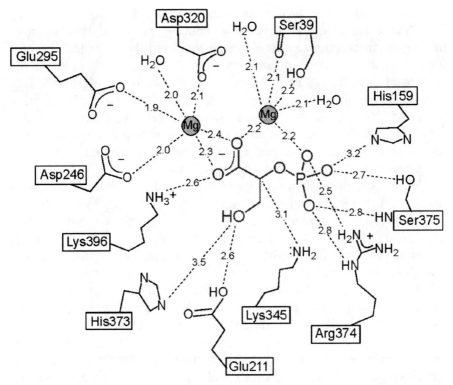

**Figure 6.2.** Structure of the active site of an enzyme (enolase) showing the bound substrate and all of the interactions between the substrate and residues in the active site of the protein. From Frey, P. A.; Hegeman, A. D. Enzymatic Reaction Mechanisms, Oxford University Press, Oxford, UK, 2007.

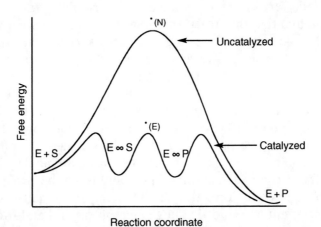

**Figure 6.3.** Reaction coordinate diagram showing the lower barrier height for the chemical conversion step (designated ‡) of the nonenzymatic reaction, ‡$_{(N)}$, versus the enzymatic reaction, ‡$_{(E)}$.

enzymes work that was put forth more than half a century ago by the Nobel laureate Linus Pauling. Pauling proposed that stronger active site interactions between an enzyme and the activated complex of its substrate, in comparison to the substrate itself, could be the source of enzyme catalysis. As shown in Figure 6.3, preferential tight binding of an activated complex brings down the height of a reaction barrier that separates the reactant from the product. Since the height of the reaction barrier determines the rate at which a reaction proceeds, this concept introduced a simple means of explaining the enormous rate accelerations that enzymes bring about.

Pauling's hypothesis held sway for many decades, forming a basis for both theoretical and experimental studies of enzymes. In fact, if you pick up your undergraduate text on biochemistry, you will find Pauling's hypothesis described in the chapter that discusses enzyme function. The past decade, however, has seen many scientists begin to question this simple picture. There are extensive pictures of enzyme active sites available from structures deposited into protein structure data banks. When we examine these static pictures carefully, we begin to question whether the changes in charge and bond distances that occur in a substrate as it undergoes conversion to its product can be the origin of a differential binding between a substrate and its activated complex to the tune of $10^{20}$-fold. Experimentally, using known three-dimensional structures for proteins, together with the tools of molecular biology, researchers have been able to alter the side chain of a single amino acid side chain far from the enzyme active site. When the ability of such a modified protein to catalyze its reaction is measured, very dramatic reductions in catalytic rate can occur, despite the fact that the structural integrity of the protein may appear completely unchanged. Experiments of this sort indicate that minor changes in a protein sequence, which do not alter the three-dimensional structure of a protein, can play an essential role in catalysis. In fact, this property of enzymes might have been inferred from numerous examples of genetically encoded diseases: when researchers map the spontaneously arising point mutations onto the three-dimensional structure of such enzymes, the changes are often found to occur all over the place and very far from the active site. While some of these changes may alter the stability of the protein, others appear to be affecting directly the chemistry of the active site.

One important aspect of Pauling's hypothesis is that, historically, it has been interpreted (too literally) in the context of static models for enzyme active sites that are derived from X-ray crystallography. But large protein molecules are by no means static in solution; instead, they are undergoing a wide range of motions taking place over femtoseconds to seconds. We refer to these motions as the "protein dynamics," and researchers now believe that a complete understanding of enzyme catalysis must include the dynamics within the protein backbone and side chains. There are many spectroscopic tools that can be applied to measuring the dynamic motions within a protein, which include fluorescence and infrared spectroscopy and nuclear magnetic resonance. I am sure you have learned about some of these methods in your introductory classes, perhaps without realizing that they can be applied to molecules the size of proteins. Theoretical chemists have also developed computational methods to examine the motions within proteins, referred to as molecular dynamics (MD) simulations. Although the timescale for such computations was originally limited to nanoseconds, it is now possible to extend the methods into longer and more catalytically relevant time regimes.

After so many years of work, extensive toolkits are now available, both to measure and simulate protein dynamics and to characterize the rates and mechanisms for enzyme-catalyzed reactions. But we are still missing an essential part of the puzzle: the "set of rules" that will allow us to link the hierarchy of motions that occurs throughout a protein to its catalytic effectiveness. This challenge remains at the forefront of research in chemical biology, in particular, with the goal of constructing a new enzyme catalyst from first principles. Quoting from a graduate student, Zac Nagel, in my lab: "While civilization has already mastered the tools to design and build the Empire State Building and the Golden Gate Bridge, we still cannot successfully engineer a new protein that is able to reproduce the enormous rate accelerations of native proteins." Perhaps with contributions from the next generation of researchers like Zac and yourself, this ultimate goal will finally be achieved.

## AH, BREATHE THAT DELICIOUS OXYGEN

I am not sure if you are a jogger—I used to be when I was younger. Now my goal is to take as long a walk as I can find time for in the morning. As a result, I spend less time gasping for oxygen, and more time contemplating the beauty of the hills near my home. Of course, a major concern these days is climate change and our impact on the environment. One thing that people rarely think about is that humans (as well as other multicellular organisms) function as little combustion machines, oxidizing reduced carbon foodstuffs into $CO_2$ and $H_2O$.

Life on planet Earth was not always like this, with the early atmosphere composed of gases such as methane, hydrogen, and $CO_2$. One of the most remarkable events to occur in the history of Earth was the evolution of photosynthetic organisms capable of using the energy of the sun to oxidize $H_2O$ to $O_2$ while capturing the released electrons to reduce $CO_2$ into biocompatible molecules. Over a period of several billions of years, these photosynthesizing organisms lived side by side with unicellular anaerobic organisms that thrived in the absence of $O_2$. The time span for aerobic organisms on Earth is actually very short (relatively speaking), corresponding to a great oxidative burst beginning about 500 million years ago. It was this event that both raised the level of $O_2$ in the environment to its current 20% and permitted the evolution of complex, multicellular organisms.

Why is $O_2$ such an important component of the evolution of complex life forms? A key feature of $O_2$ is the enormous energy released upon the four-electron reduction of $O_2$ to water, $E° \cong 1 V$ and $\Delta G° \cong -23 kcal/mol$. Yet, aerobic organisms manage to live very well in an $O_2$-containing atmosphere. This is because the prevalent spin state of $O_2$ is as a ground-state triplet, while the majority of organic molecules are ground-state singlets; that is, their direct interaction is a spin-forbidden process. If the dominant form of $O_2$ in the atmosphere was instead a ground-state singlet, $O_2$ would be undergoing explosive reactions with carbon-based life forms. It turns out that the defining property of ground-state triplet $O_2$ is its kinetic sluggishness coupled to the huge energy released upon its reduction to water.

How have biological systems learned to handle this dichotomous behavior of $O_2$? Once again, enzymes come to the rescue, with their ability to manage simultaneously the slow reactivity of $O_2$ and its potential thermodynamic explosiveness. The key to the success of these enzymes is their catalysis of a controlled, stepwise addition of electrons to $O_2$, according to the cycle of Figure 6.4. From a catalysis perspective, the pathway in Figure 6.4 creates transient, $O_2$-derived intermediates that possess the same spin state as the substrate that undergoes oxidation, allowing reaction to proceed on timescales compatible with life. In order to perform this type of stepwise reduction, enzymes have often paired with special classes of small inorganic and/or organic cofactors.

If you take a minute, Angela, and inspect the partially reduced intermediates (Figure 6.4) that lie between $O_2$ and water, you may be surprised to see species that are normally considered to be toxic to

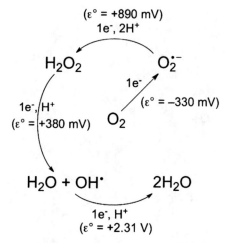

**Figure 6.4.** *Cycle for the reactivity of $O_2$ showing the intermediates derived from a stepwise addition of electrons and protons together with the redox potentials for these steps. A negative potential means it is uphill, and a positive potential means it is downhill. From Klinman (2007).*

the cell, in particular, the superoxide ion ($O_2^{\cdot-}$) and the hydroxyl radical (HO·). If an enzyme must generate these species in order to utilize the energy content stored within the $O_2$ molecule, how does it manage to escape radical damage to itself or to other components of the cell? One strategy is to take advantage of a metal ion that can both transfer electrons to the $O_2$ and bind/restrain the resulting reduced $O_2$ intermediate. This is a very general theme among metalloenzymes containing a single metal ion, multiple metal ions, or a metal ion complexed with an organic cofactor.

One of the most extensively studied enzymes of the latter class is cytochrome $P_{450}$, an important enzyme in clearing drugs and foreign compounds (referred to as xenobiotics) from the body. Cytochrome $P_{450}$s contain a heme–iron complex similar to the cofactor in hemoglobin that carries $O_2$ in the blood (Figure 6.5). One important difference between cytochrome $P_{450}$ and hemoglobin is the nature of the amino acid side chain that anchors the cofactor into the enzyme active site: cysteine in cytochrome $P_{450}$ as opposed to histidine in hemoglobin. This relatively simple change makes it possible for both the iron and the heme in cytochrome $P_{450}$ to donate a total of three electrons into $O_2$ (with a fourth electron coming from an outside source). In this manner, cytochrome $P_{450}$ has figured out how to donate the four electrons normally required for the complete reduction of $O_2$ to water to generate, instead, a highly reactive but sequestered species able to oxidize most hydrocarbons.

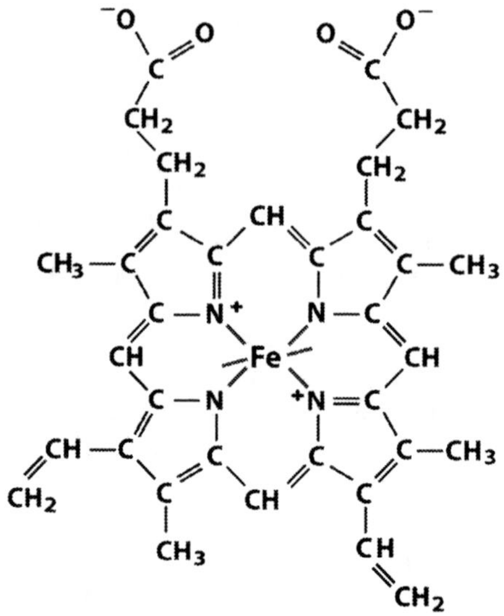

**Figure 6.5.** *A two-dimensional picture of the tetrapyrrole at the active site of the $O_2$ carrier protein, hemoglobin. There are six bonds to the Fe: four of these come from the tetrapyrrole and two from the additional ligand, one above and one below the plane of the paper. Source: Nelson, D. L. and Cox, M. M. Lehninger's Principles of Biochemistry, 5th Ed.*

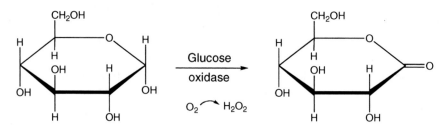

**Figure 6.6.** *Reaction catalyzed by the enzyme, glucose oxidase, in which glucose is converted to a gluconolactone and $O_2$ is converted to hydrogen peroxide.*

By contrast, it would appear that the utilization of $O_2$ may become particularly dire when an enzyme lacks any means of securely anchoring the reduced intermediates that arise from $O_2$. Glucose oxidase, a much-studied enzyme that uses an organic cofactor (a flavin) to catalyze the oxidation of glucose to a gluconolactone together with the reduction of $O_2$ to hydrogen peroxide, offers some insight into this vexing problem (Figure 6.6). Glucose oxidase has many practical applications, for example, as a biosensor to monitor levels of glucose in the blood, making it an attractive target for intensive study. During its catalyzed reaction, the flavin of glucose oxidase first receives two electrons from the glucose substrate and then delivers these *one at a*

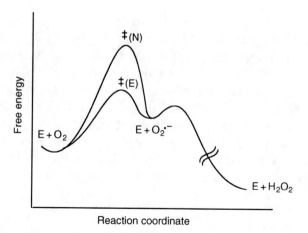

**Figure 6.7.** *Free energy diagram showing how an enzyme can lower the barrier for $O_2$ reduction without changing the energy of the reduced $O_2$ intermediate. The nonenzymatic transition state is ‡$_{(N)}$, while the enzymatic transition state is ‡$_{(E)}$.*

*time* to $O_2$ to form an intermediate superoxide ion. Superoxide is actually a potent reductant; that is, it has a high tendency to *give up* its electron. One strategy the enzyme could use to facilitate its reaction would be to stabilize the superoxide intermediate. However, this would lead to the accumulation of a potentially damaging intermediate at the enzyme site. When the properties of the reaction of $O_2$ with glucose oxidase were studied, it was found that a single positively charged amino acid side chain near the bound flavin cofactor played a key role in the catalysis. Most importantly, the enzyme has "figured out" how to use this point charge to increase the rate of formation of the superoxide without altering its stability. The rapid donation of a second electron to produce hydrogen peroxide further ensures that the lifetime of the enzyme-bound superoxide ion is very short. This course of action can be illustrated very simply by using a free energy diagram (Figure 6.7), which illustrates the specific impact of glucose oxidase on the kinetic barrier for $O_2$ reactivity.

## BYE FOR NOW ...

Well, Angela, I hope I have convinced you that enzymes are some of the most fascinating molecules on the planet. Not only have they evolved into incredibly powerful catalysts with rates of reaction that can be as much as $10^{20}$-fold greater than for comparable reaction in solution but they have also managed to provide the foundation for the existence of complex life in a potentially toxic ($O_2$) gaseous environment. The future holds much excitement, with possibilities of designing new enzyme catalysts from first principles, the creation of new materials that are a hybrid of biological and synthetic molecules, and the exploitation of enzymes to reduce carbon emissions and to help us stem the urgent problem of increased global warming. I do hope I have captured your imagination! If you take a trip to Northern California, stop by and I will give you a tour of the lab and introduce you to the graduate students and postdoctoral fellows who are making that future a reality.

Sincerely,

Judith

## FURTHER READING

Klinman, J. P. How do enzymes activate oxygen without inactivating themselves? *Accounts of Chemical Research* 2007, *40*, 325–333.

Nagel, Z.; Klinman, J. P. Tunneling and dynamics in enzymatic hydride transfer. *Chemical Reviews* 2006, *106*, 3095–3118.

# 7

# *Making Sense of Oxygen*

**Marie-Alda Gilles-Gonzalez**
*UT Southwestern Medical Center at Dallas*

Dr. Gilles-Gonzalez was born in Jeremie, Haiti. She earned a BS degree in biochemistry from State University of New York at Stony Brook and a PhD in biochemistry under Dr. Gobind Khorana at MIT. Since completing her postdoctoral studies with Dr. Max Perutz at the MRC Laboratory of Molecular Biology in England in 1995, she has directed her own laboratories, first at the Ohio State University and now at the UT Southwestern Medical Center, where she and her colleagues study biological oxygen sensors.

Dear Angela,

So you do not wear unmatched socks or untied shoelaces. You comb your hair and, my word, you look exotic and even wear a little make up! And you have the temerity to aspire to be a chemist. Good! You appear not to be a careerist. I have no word for those who attempt to recruit me to assist them in mapping out every twist and turn of their lives. One can hardly be expected to serve as a "role model" (How I hate that expression!) when the lives of scientists are so serendipitous. I do assume you to be a scientist: a young one, in the

---

*Letters to a Young Chemist*, First Edition. Edited by Abhik Ghosh.
© 2011 John Wiley & Sons, Inc. Published 2011 by John Wiley & Sons, Inc.

rough. If so, I cannot beat you away from this path, though it is assuredly not for the faint of heart. My scientific heroes are Galileo Galilei and Rachel Carson: the sort who favored observation over the rhetoric of their times and paid dearly for this. Who are yours? A good friend, now famous and retired, tells me that his minister father held a special prayer for his soul before letting him go on to graduate school. Perhaps his father was right. My friend is a well-scarred veteran of many scientific battles. Despite his many awards and titles, I can reliably get his dander up by bringing up a name or two. "*E pur si muove.*" Indeed.

I wish I could say that your looks and social background will not matter in a line of work as dispassionate and objective as chemistry. How could they not? Science is an enterprise done in society. If in the United States research in chemistry is currently dominated by one demographic group that looks nothing like you, you should take this as sign of poor health and an indication that there is a crying need for people like you. Indeed, therein lies your advantage; since no one has told you how things should work, you might just find a new way. I admit that I am not entirely without self interest in encouraging you on this tentative step. My scientific ambitions are much larger than the acquisition of personal fame. I dream of a time when, for example, the science of African Americans will be recognized to be as excellent and distinctive as jazz. Think of Johnny Hartman's voice and John Coltrane's saxophone doing Billy Strayhorn's "Lush Life," with McCoy Tyner on piano, Jimmy Garrison on bass, and Elvin Jones on drums. Close your eyes and hold that thought.

## RETRACING MY ROOTS

Like you, I am told I grew up poor, though I cannot honestly claim I did. True, my childhood was spent on the island of Haiti crammed with my extended family into one or two small rooms of decrepit colonial houses. To see this, however, as the only notable aspect of my childhood would hardly do justice to the color and magic that pervaded my daily life, or the people and spirits who populated it. How might a passing tourist or Red Cross worker ever learn of the days spent inside, rapt by the adults' embellishments on our many folk tales, while the furious hurricanes laid low our tall palms. How could one have been poor where material wealth was never a concern? I grew to know that my relatives sacrificed daily to support my well-being, and the benefits of such a loving, extended family implied some debt of responsibility. Initially, I thought my part in this would be to provide material support. Lacking the habit of a middle-class upbringing to channel every instinct into a possible career, I nearly skipped college altogether, and when I did enter a university, I did so thinking I would train to become a medical doctor. The motivation for this choice could not have been more trivial: a medical doctor was the person with the highest level of formal, technical education whom I had encountered, and I was considered a talented student who would "go far." Where to? No one really knew, least of all me. Still, I was not permitted to meander there. The notion of wandering about and "finding oneself," then a highly fashionable idea, was not an option. My family members, all of whom lacked the benefits of a formal education, were smart, admirable, and highly resourceful people, and I had been charged with the intimidating task of bettering them (Figure 7.1). Fortunately, a university offered an almost infinite web of choices. There, my lack of direction became an asset. Organic chemistry, with its clarity and logic, drew me to itself like no other subject.

Do apply yourself well to your studies of organic chemistry, and with this, physical chemistry and mathematics. One must master these subjects to understand biology in any way that is more than descriptive or superficial. One must respect biology too, if living organisms

*Figure 7.1.* The author in Haiti with the women of her family. The photograph shows (counterclockwise from bottom left) the author at age eight with her great-grandmother, mother, grandmother, and aunts.

are what one seeks to understand. Unsubstantiated flights of fancy into the supposed chemistry of living things can definitely obscure rather than clarify the paths toward solving biochemical problems.

## OXYGEN SENSORS

Consider, for example, how living beings sense molecular oxygen ($O_2$). Admittedly, this problem is my own personal passion. "Why should living things sense $O_2$?" you might ask. Indeed that question goes to the heart of the problem. The earth's atmosphere currently contains 20% $O_2$ (the rest being almost entirely $N_2$). This was not always so. All of the $O_2$ in the atmosphere resulted from the actions of living organisms. This gas first began to accumulate about 2.5 billion years ago, when cyanobacteria learned to split water to $H_2$ and $O_2$, as a way of harvesting the energy of the sun. Back then, living things had not yet learned how to handle $O_2$ safely, and being an oxidizing and caustic gas, it was a poison to most. What recourse could there have been for the other microorganisms? One can only surmise that most of them dug themselves in as deeply as they could, and the more ingenious ones among them developed ways to sense the $O_2$ so as to avoid it. Today, one can still find ancient microorganisms, such as *Methanobacterium thermoautotrophicum* and *Aeropyrum pernix*, that still use their $O_2$ sensors to avoid $O_2$. These sensors bind $O_2$ with a heme, a molecule you have no doubt encountered in lessons about hemoglobins. Indeed, sensors such as the one in *A. pernix* are the heme protein ancestors of the modern myoglobins and hemoglobins that today serve to transport $O_2$. In other words, living creatures began to sense $O_2$ because it was a poison to them, and they needed to avoid it. Though $O_2$ is still a poison, as Paracelsus has said, "The right dose differentiates a poison from a remedy." We and other living creatures have not only raised the allowable dose of this poison but have also learned to depend on it so as to burn our foods and to extract energy from them.

As far as we know, most living bacteria that sense $O_2$ nowadays do so to gain information about their position rather than to avoid $O_2$. Consider, for example, a jar of pure water that is bubbled with air at room temperature (23°C) and normal atmospheric pressure. After this water is completely equilibrated with air, it will have about 250 μM $O_2$. This concentration of $O_2$ can also be expected of anything that is completely equilibrated with air; however, few living cells ever achieve such equilibration. In our own bodies, the surface layer of cells in our lungs is the only one that benefits from such a high concentration of $O_2$. A few layers of cells beyond this, the concentration of $O_2$ dramatically drops to only a few percent of air saturation. Something even more drastic happens for microorganisms such as bacteria. In nature, bacteria live mostly in biofilms, which are very thin films formed from their own excretions of large molecules (mostly polysaccharides) onto solid substrates. For *Pseudomonas aeruginosa* or *Stapholococcus aureus* bacteria, these films are merely 0.2 mm thick, and the concentration of $O_2$ drops from 250 μM to nearly 0 μM as one moves across a thickness as short as 0.05 mm. This very steep drop in $O_2$ happens because the living bacteria are actively consuming this gas. Regardless of how this hypoxia comes about, the bacteria in the different layers of a 0.05-mm-thick biofilm are living in environments that differ vastly more in their $O_2$ concentrations than sea level versus the top of Mount Everest would feel to a human lung (e.g., about 256 μM vs. 85 μM $O_2$).

## THE FIXL SENSOR

Incidentally, though it is now clear that pathogenic bacteria gain a high degree of protection from antibiotics while living in biofilms, and their $O_2$ sensors will undoubtedly be important to medicine, work motivated by a desire to understand sensors, rather than the wish to generate new drugs, is what has led to a greater understanding of those sensors. In fact, the foundational work on bacterial $O_2$ sensors was done not on pathogenic bacteria but on the beneficial rhizobial species. Rhizobia live symbiotically with leguminous plants in root nodules, where they fix nitrogen gas from the atmosphere into fertilizing nitrogenous salts for the plants. These bacteria can only carry out this nitrogen fixation chemistry in the complete absence of $O_2$. They use their $O_2$ sensors to learn whether or not they are inside a root nodule that the plant has scrubbed of $O_2$ but has abundantly fueled with adenosine triphosphate (ATP).

These $O_2$ sensor proteins discovered in rhizobia are kinases called FixL. Like many proteins that serve in biological signaling (or signal transduction), these proteins "translate" information about an external signal into chemical language that a cell can readily understand. Using a combination of chemistry and biology, my colleagues and I have managed to show that FixL conveys information about the absence of $O_2$ from its heme cofactor by phosphorylating a transcription factor called FixJ (also a protein) in the following reaction:

$$ATP + FixJ\text{---}deoxy\text{-}FixL \rightarrow ADP + p\text{-}FixJ$$

The p-FixJ then goes on to activate all of the genes in the rhizobia that are necessary for fixing nitrogen.

The FixL protein has a beautiful dark red color, much like blood, when it is exposed to air (Figure 7.2). When air is removed, FixL acquires a slight purple tinge. Alternatively, it becomes a bright cherry red when carbon monoxide is added to it. Such changes in the way

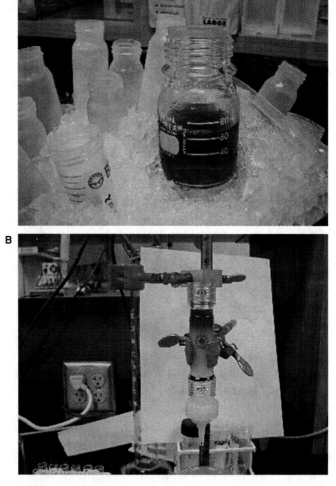

**Figure 7.2.** Purification of an oxygen sensor protein. (A) A solution of the protein Escherichia coli direct oxygen-sensing cyclase (Ec DosC) after it is partially purified by precipitation from a solution of salt (ammonium sulfate). (B) The same protein adsorbing onto a column packed with an anion-exchange matrix (diethylaminoethyl sepharose); at the neutral pH being used here, the matrix is retaining the protein because it carries a net negative charge.

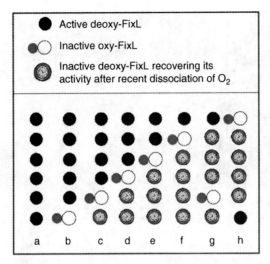

**Figure 7.3.** *Schematic representation of the memory effect in FixL. Oxygen (small dot) rapidly associates with deoxy-FixL (filled circle) (a) and immediately switches off the kinase (open circle) (b). After oxygen dissociates from the FixL (c), the protein only slowly recovers its activity (clock). Therefore, a single molecule of oxygen can switch off multiple deoxy-FixLs (c–e). Since the time it takes the protein to recover its kinase activity is longer than the average time it takes a deoxy-FixL to find an oxygen molecule, many recovering deoxy-FixLs will rebind oxygen before they can fully regain ther activity (f, g). Ultimately, an equilibrium is reached; the proportion of FixL molecules in different states will no longer change, but different molecules alternate between these states (f–h).*

a protein absorbs light are quite fortunate and can be exploited to calculate how fast and how well $O_2$, or any other molecule, interacts with a heme-containing protein like FixL. Recently, we managed for the first time to examine the dose response of FixL, that is, how well FixL forms p-FixJ as we add to it more and more $O_2$. This study required combining information about $O_2$ binding to FixL together with information about FixL's rate of conversion of FixJ to p-FixJ. It was gratifying to see that our results in a test tube completely agreed with the general response of rhizobial nitrogen fixation to $O_2$. Importantly, careful quantification of our data helped us to discover that FixL can actually remember having bound $O_2$ (Figure 7.3).

## IT'S ALL CONNECTED

As the great environmentalist John Muir wrote, "When one tugs at a single thing in nature, he finds it attached to the rest of the world."

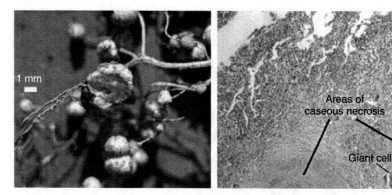

**Figure 7.4.** Beneficial and pathogenic interactions of bacteria with their hosts are gauged by analogous oxygen sensors. The left panel shows a root nodule where a symbiotic interaction takes place between the beneficial bacteria Bradyrhizobium japonicum and their soybean host. When a healthy nodule is sliced, it looks pink because a heme protein called leghemoglobin is made by the plant to keep down the level of free oxygen, and this correlates with active nitrogen fixation by the bacterial symbiont. The right panel shows a small area of an M. tuberculosis–infected human lung; areas with the latent pathogen are characterized by granuloma with fibrous margins and areas with caseous (cheeselike) necrosis in their center. The granuloma and giant cells represent an attempt by the body to wall off the infectious agent.

My work on FixL has led me to suspect that in science, it might not matter at all what one studies, so long as one studies it well. For example, my colleagues and I recently learned that two relatives of FixL exist in *Mycobacterium tuberculosis* (Mtb), a pathogen that currently infects one-third of all humans. Fortunately, many of these infections are latent, with the bacteria sometimes remaining dormant for decades. In Mtb, there are two FixL-like sensor proteins called DevS and DosT. Just like FixL, DevS and DosT phosphorylate a transcription factor that activates many genes. In Mtb, however, this represents not a cooperative arrangement but an urgent response to a human immune system that is determined to suffocate it (Figure 7.4). As a result, a truce is reached. In this new arrangement, the bacteria survive, but they cannot thrive—at least not while the human immune system remains healthy.

We have discovered too that some relatives of FixL are able to translate information about $O_2$ binding into different types of chemistries. Examples of this are a class of cyclase enzymes (called GGDEF domain cyclases) that synthesize cyclic nucleotides and a set of phosphodiesterase enzymes (called EAL domain phosphodiesterases) that can cleave these cyclic nucleotides. These enzymes, which are inti-

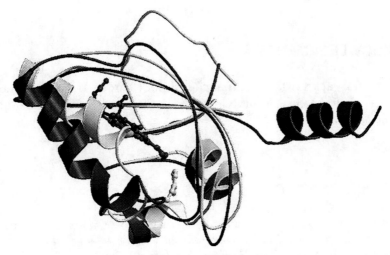

**Figure 7.5.** *Comparison of the structure of an oxygen detector to that of a light detector. Here, a cartoon structure of the heme-binding region of* B. japonicum *FixL (dark structure) is compared with that of the light-detecting region of the photoactive yellow protein (pale structure). The dark molecule in the center of the FixL structure is the heme; it occurs only in FixL and directly binds to the oxygen. The pale molecule at the bottom of the image is a hydroxycinnamate cofactor that occurs only in the photoactive yellow protein and directly detects light.*

mately involved in biofilm formation, are in many cases probably important to the damaging actions of numerous pathogenic bacteria.

Finally, I shall leave you to chew on the thought that the $O_2$-detecting portion of FixL very much resembles the light-detecting part of a class of bacterial light sensors (Figure 7.5). Thus, it seems that nature, by a combination of great patience and careful selection, managed to fabricate an $O_2$ sensor and a light sensor from the same mold. Could an engineer ever pull off such a feat? Discoveries such as these, which can tug on important aspects of the natural world and bring them to light, are an unending source of wonder—that and the company of good scientist friends with whom to share this great pleasure.

Good luck with your studies.

All the best,

Marie-Alda

## ACKNOWLEDGMENTS

The author acknowledges financial support from the US National Science Foundation Grant No. MCB620531 and Welch Foundation Grant No. I-1575.

## FURTHER READING

Gilles-Gonzalez, M. A.; Gonzalez, G. A surfeit of heme-based sensors. In *The Smallest Biomolecules: Diatomics and Their Interactions with Heme Proteins*, Ghosh, A. (ed.), Elsevier, Amsterdam, 2008.

Lane, N. *Oxygen: The Molecule That Made the World*, Oxford University Press, Oxford, UK, 2002.

# 8

# *Let's Visualize Biology: Chemistry and Cellular Imaging*

**Elizabeth M. Nolan**

*Massachusetts Institute of Technology*

Elizabeth M. Nolan is an assistant professor of chemistry at the Massachusetts Institute of Technology (MIT). She received her BA from Smith College (Northampton, Massachusetts), where she did research with Professor R. G. Linck on stereoelectronic effects in simple substituted alkanes. She earned her PhD from MIT under the mentorship of Professor Stephen J. Lippard, where her research addressed the design, characterization, and use of small-molecule fluorescent sensors for imaging zinc in neurobiology and for detecting mercury in aqueous solution. She then pursued postdoctoral research under the guidance of Professor Christopher T. Walsh at Harvard Medical School, where she studied the biosynthesis of an antimicrobial peptide–siderophore conjugate microcin E492m.

*Letters to a Young Chemist*, First Edition. Edited by Abhik Ghosh.
© 2011 John Wiley & Sons, Inc. Published 2011 by John Wiley & Sons, Inc.

Hi Angela,

How are you doing? Are you enjoying your sophomore year, your classes, and your new dorm? I'm sorry that we missed you at the annual family reunion last week! But I understand how busy the start of the semester can be, and air travel is no fun these days with so many flight cancelations and delays (and your trip home from Norway wasn't so long ago!). I'm sure that your mom filled you in about the party and all of the news from our relatives. Needless to say, it was a fun gathering and I hope that you'll make it next year. Uncle Frank mentioned that he was in San Diego recently and that you both caught up and discussed your future plans. I was happy to learn that you're thinking about pursuing chemistry, and I can't wait to hear about the summer research you did in Norway. Uncle Frank suggested that I share a bit about my experiences in graduate school and as a postdoc with you. Subsequently, Aunt Martha mentioned that you expressed interest in learning more about how chemists can study biological systems, and she thought that you might like to hear a bit about how chemists contribute to biological imaging. Aunt Martha also said that you fell off your bike, hurt your knee, and needed a magnetic resonance imaging (MRI) scan right before your Norway trip—what a relief that everything ended up being OK! I wonder if getting an MRI sparked your interest in imaging. ...

My postdoctoral research in enzymology and natural product biosynthesis was pretty far removed from the imaging field, but my PhD thesis work was in this area. During grad school, I aimed to make small-molecule fluorescent sensors that could be used to visualize metal ions in cells or biological tissues. You might ask, why small molecules? What does "fluorescent" mean? How do you go about doing this type of research? Hopefully you'll have a better understanding by the time I'm through writing to you. In what follows, I'll tell you some about small-molecule fluorescent sensors like the ones I made as a grad student; however, I think it's most important to provide you with an appreciation of the diversity of strategies and approaches for biological imaging.

Broadly, this field involves the use of spectroscopy (i.e., fluorescence and magnetic resonance) or radioactivity to image biological structures and phenomena. Chemists from all subspecialties can contribute to imaging! Advances in imaging require new instruments, new molecules, and new analytical methods. Furthermore, imaging provides opportunities for terrific collaborations between chemists, scientists in other disciplines (physics, biology, etc.), and even clinicians. During my PhD, I was lucky to work closely with some neurobiologists. I learned so much; the day-to-day laboratory work was very different, and I realized that there are real advantages to having chemists and biologists work together. Maybe you'll be involved in a similar collaboration someday! Anyway, before I get sidetracked with ruminations and stories, my background is in inorganic chemistry and biochemistry, so I want to tell you about how chemists contribute to cellular imaging by providing new "tools," generally proteins and synthetic small molecules, and methodologies for visualizing cells, organs, biological phenomena (enzymatic activity, protein translocation, signaling, etc.), and analytes within cells or tissue samples. If you're curious to learn more about how new spectroscopic approaches are developed and used or instrument design, just let me know and I can point you toward some relevant literature.

## FLUORESCENCE AND FLUORESCENT PROTEINS FOR IMAGING

I'll begin by telling you a bit about fluorescent proteins and how they've revolutionized experimental biology. A key contributor to this area is Professor Roger Y. Tsien of University of California, San Diego (UCSD)! As I'm sure you know, Professor Tsien, along with Martin Chalfie and Osamu Shimomura, received the 2008 Nobel Prize in Chemistry for the discovery and development of a fluorescent protein called green fluorescent protein (GFP). You should definitely look up his Nobel Prize lecture and maybe even make an effort to visit his laboratory.

First, you might be wondering, what is fluorescence and why is it useful for cellular imaging? Fluorescence occurs after a molecule absorbs energy in the form of light (recall from high school chemistry that $E = h\nu$, where $E$ is energy, $h$ is Planck's constant, and $\nu$ is the frequency of the light), which results in the excitation of one of its electrons to a higher energy state called an "excited state" (Figure 8.1). There are many excited states and, after the initial excitation, the electron will undergo "nonradiative decay," which means that the electron will lose some energy and fall to an excited state with lower energy than the initial excited state. After some period, the electron will decay back to the ground state and release a photon of energy

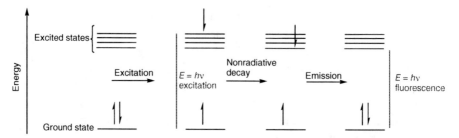

**Figure 8.1.** *Simple energy level diagram for the excitation of an electron from the ground state of a fluorophore, nonradiative decay of the electron to an excited state of lower energy, and return of the electron to the ground state, which results in the observed fluorescent light (emission).*

in the form of light. This light is fluorescence or "emission" (a small caveat here is that the spin of the excited electron must be the same in the ground and excited state for fluorescence to occur, but I won't go into these details right now). A fluorescence or emission spectrum looks like a concave-down curve, similar to an absorption spectrum. Maybe you collected absorption spectra during general chemical laboratory and used the $A_{max}$ values in a Beer's law plot. The excitation and emission wavelengths of a molecule generally correspond to the wavelengths of maximum excitation and emission. As a result of nonradiative decay in the excited state, the fluorescence emission spectrum is generally shifted to longer wavelengths (lower energy because $E = hc/\lambda$, where $c$ is the speed of light and $\lambda$ is the wavelength) relative to the corresponding excitation spectrum. In other words, the excitation maximum is generally of higher energy (shorter wavelength) than the emission maximum. The difference between the excitation and emission maxima is called the "Stoke's shift."

A fluorophore is a component of a protein or small molecule that exhibits fluorescence. It can also be called a "fluorescent label," "chromophore," or "fluorescent probe." Each fluorophore has characteristic excitation and emission wavelengths. In other words, a fluorophore will fluoresce when the light of a particular energy, corresponding to the excitation wavelength, is used. Examples of fluorophores include GFP and related emissive proteins, and small molecules like fluorescein and coumarin. Many biomolecules have intrinsic fluorescence. For instance, tryptophan, an amino acid, fluoresces in the ultraviolet region of the electromagnetic spectrum.

Some fluorophores are used only to label a given organelle or protein. Others are "reporters" or "sensors" because they are engineered to exhibit a fluorescence change following some perturbation or analyte recognition. For instance, some sensors provide a change in fluorescence intensity, either "turn-off" or "turn-on," with analyte recognition (think about switching a light bulb on and off). Other sensors are "ratiometric" because a ratioable change in excitation and/or emission wavelength occurs. In other words, the excitation or emission spectrum of the ratiometric sensor shifts to either lower or higher wavelengths with analyte recognition. Fluorescent sensors have been used to detect/image enzymatic activities, metal ion (Ca, Zn, Cu, etc.) distributions, and the presence of small molecules like nitric oxide (NO) and hydrogen peroxide ($H_2O_2$) in live cells.

Fluorescence microscopes can be used to measure and monitor fluorescence inside of cells or in tissue samples. Just like computers or cell phones, there are many different types with different features. Describing all of the possibilities would take eons! To provide some examples, epifluorescence microscopes utilize a xenon lamp as a light source for excitation, whereas confocal microscopes employ a laser system to excite the sample. Under ideal circumstances, a fluorescence microscope will have a "bright field" or "phase contrast," which allows a picture of the sample to be taken when illuminated with white light. This image provides information about the spatial orientation of the cells or tissue sample in the petri dish. Then, the microscope can be switched to fluorescence mode and the fluorescent regions of the sample can be identified and studied. Sometimes, for "live imaging" studies (the cells are living, as opposed to looking at "fixed" cells that are dead and attached to a microscope slide), it is necessary to have the microscope inside of an incubator to keep the temperature and carbon dioxide levels suitable for cell survival. These systems can be very expensive and time-consuming to maintain. Although most laboratories that do substantial imaging have their own microscope(s), most universities maintain microscopy facilities that can be used by students and postdocs on a sign-out basis. You might want to see if UCSD has such a facility. Visiting one and having a brief tour would be very informative!

Anyway, let's get back to fluorescent proteins and their uses in cellular imaging. The bottom line is that fluorescent proteins have revolutionized biology! If you take a few moments to flip through journals like *Cell*, *Science*, and *Neuron*, you're bound to find multiple papers where fluorescent proteins were employed for studies of all sorts of cellular phenomena. They might be used to label an organelle, to monitor receptor trafficking, or to visualize protein synthesis. My guess is that you've already heard a lot about GFP, which was isolated from the jellyfish *Aequorea victoria*. This protein is called

GFP because it has an emission wavelength at 509 nm, which is in the green region of the visible spectrum. As a result, it emits green light following excitation (its major excitation peak is at 395 nm). The protein has a beta barrel structure and the fluorophore is located in the center of the barrel. After the discovery and cloning of GFP, a technique called mutagenesis was employed to make numerous GFP variants. In mutagenesis, which you might learn about in a molecular biology course, changes to the DNA sequence are made in order to substitute an endogenous amino acid with an alternative one. Through this approach, GFP variants with different fluorescence properties (greater brightness, different colors, etc.) were developed. Many of these variants are used daily in laboratories across the world. Some examples include enhanced green fluorescent protein (eGFP), yellow fluorescent protein (YFP), and cyan fluorescent protein (CFP).

Why are GFP and its derivatives such powerful tools for biology and cellular imaging? The answer to this question lies at the heart of molecular biology. It's very easy to attach GFP to other proteins or peptides using molecular biology techniques (I'll skip the details for now, but the techniques fall into the category of "cloning") and then introduce and express these GFP–protein conjugates into cell cultures. For instance, when GFP is attached to a protein/peptide that contains an intracellular localization signal (generally a sequence of amino acids), the GFP can be directed to an intracellular target like the mitochondria or Golgi apparatus. This technique allows for precise control over GFP localization. If a mitochondrial directing sequence is attached to GFP, the protein, and hence GFP fluorescence, will only be present in the mitochondria. If GFP is fused to a membrane protein, its fluorescence will only be seen where this protein expresses in the cellular membrane. In my PhD work, I used fusion proteins of red fluorescent protein (RFP) to label cellular organelles like the Golgi apparatus and mitochondria for imaging studies. We needed to mark these organelles because we aimed to determine the cellular localization of some zinc sensors that I synthesized. Because the zinc sensors provided green fluorescence, we could compare and overlay the red from the RFP and the green from the zinc sensor to determine its localization. Because mixtures of red and green make yellow, an overlay that shows yellow indicates good colocalization of the fluorescent probes under study.

In addition to their use as labels or tags, fluorescent proteins can also be used in sensors. One approach is to create a fluorescence resonance energy transfer (FRET) reporter. FRET reporters must contain two fluorophores with different spectral characteristics. These fluorophore combinations are often called "FRET pairs," and examples include fluorescein/rhodamine and CFP/YFP. In most FRET reporters, the two fluorophores are separated by the analyte recognition element. For instance, CFP and YFP might be separated by another protein or a peptide sequence that binds metal or can be modified by a particular type of enzyme. In the absence of an analyte (i.e., metal, enzyme), the fluorophores are far apart and do not interact. Excitation of CFP provides CFP emission, and likewise, excitation of YFP provides YFP emission. In the presence of an analyte, the conformation of the FRET reporter changes and the two fluorophores come closer together. As a result, excitation of CFP results in energy transfer to YFP (the excited CFP electron essentially jumps to the YFP excited state) rather than CFP emission. Because the excited CFP electron jumped to the YFP excited state, emission from YFP is observed instead of CFP emission. A comparison of CFP and YFP emission before and after analyte recognition provides a ratiometric measurement of analyte concentration. Like GFP fusion proteins, a FRET reporter can be expressed in living cells and used to monitor intracellular phenomena. Some examples include the use of FRET to monitor intracellular calcium flux (some of these FRET reporters, developed by the Tsien laboratory at UCSD, are called "chameleons"), zinc entry into the mitochondria, and enzymatic activity.

## FLUORESCENT SMALL MOLECULES FOR IMAGING

Remember that I mentioned my PhD work involved the synthesis of fluorescent small molecules? Fluorescent small molecules (Figure 8.2) are alternatives to fluorescent proteins for *in vivo* imaging. Some argue that small molecules, like fluorescein and rhodamine, offer advantages over proteins because they are much smaller than fluorescent proteins and therefore should cause less perturbation to the native biological system under investigation.

Before I forget, one great way to get an idea of the types of small molecules used in cellular imaging is to get an old copy of the *Molecular Probes* (now a part of the company Invitrogen) catalog and flip through the pages (Haugland 2002). A tremendous array of fluorophores is available commercially, and the catalog provides helpful diagrams of their structures and excitation/emission profiles. Some of the dyes have funny names like Oregon Green and Texas

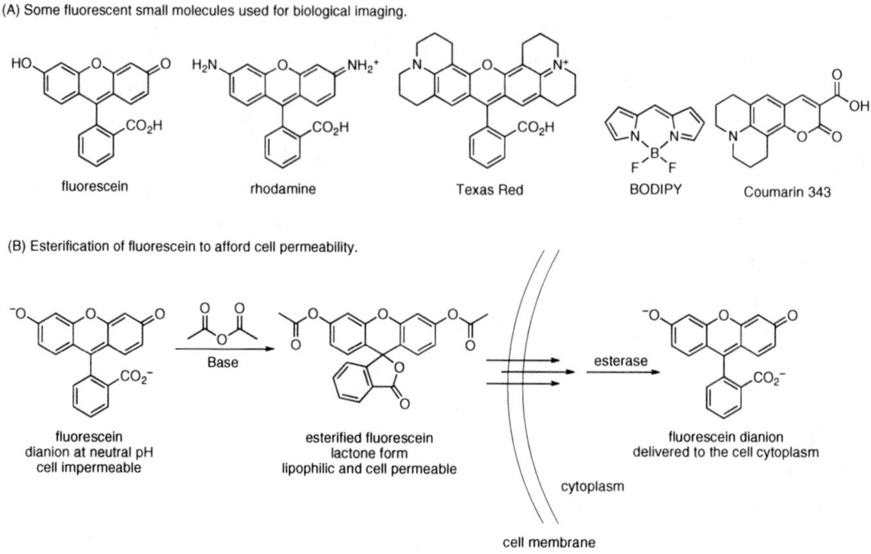

(A) Some fluorescent small molecules used for biological imaging.

| fluorescein | rhodamine | Texas Red | BODIPY | Coumarin 343 |

(B) Esterification of fluorescein to afford cell permeability.

| fluorescein dianion at neutral pH cell impermeable | esterified fluorescein lactone form lipophilic and cell permeable | fluorescein dianion delivered to the cell cytoplasm |

**Figure 8.2.** *(A) Some examples of small-molecule fluorophores used for biology and cellular imaging studies. These fluorophores can be modified synthetically to provide fluorescent molecules with properties like biological analyte sensing capabilities. (B) Chemical reaction that affords an esterified version of fluorescein and subsequent hydrolysis of the ester linkages by intracellular esterases following entry into the cytoplasm.*

Red. The names don't tell you much about the chemical structures but at least indicate the type of emission. Both Oregon Green and Texas Red are xanthenone-based chromophores. Oregon Green is a fluorescein derivative. Fluorescein and its derivatives are commonly employed in biology because they are water compatible, require visible excitation (higher energy excitation, like ultraviolet, is damaging to cells, just like how getting a sunburn is damaging to your skin!), and are very bright at physiological pH (some fluorophores have pH-sensitive emission). Texas Red is a rhodamine derivative. Boron dipyrromethene (BODIPY) is another small-molecule fluorophore that is also used in cellular imaging; it has a very high extinction coefficient (recall Beer's law: absorption $= \varepsilon \times l \times c$), which contributes to its bright fluorescence. Other small molecules used for imaging include coumarin, julolidine, and seminapthofluoresceins. Many other fluorophores exist (pyrene, quinoline, etc.), but some of these lack the water solubility or bright emission that is an ideal characteristic for many biological experiments.

Synthetically, all sorts of modifications can be made to fluorophores. These changes can result in molecules with different properties like cellular distribution or reactivity. One important example involves modification of the xanthenone fragment of the fluorescein. Fluorescein, despite all of its use for intracellular imaging cannot permeate the cell membrane! The trick is to esterify the xanthenone moiety (its protection group chemistry, like what you'll study in organic chemistry this year), which results in a neutral and much more lipophilic form (Figure 8.2). A lipophilic compound has an affinity for lipids, and the cell membrane is a lipid bilayer. As a result of esterification, the fluorescein can cross the cell membrane. Once inside the cell, enzymes called esterases hydrolyze the ester bonds and unmodified fluroescein is released. These simple reactions of ester bond formation and hydrolysis provide stealth means to deliver an otherwise impermeable molecule into a cell.

If you look in the *Molecular Probes* catalog, you'll see that a number of small-molecule fluorophores are sold as organelle markers. For instance, MitoTracker Red is sequestered by the mitochondria and LysoTracker Green localizes to the lysosomes. Hoescht is a stain that is specific for cell nuclei. It provides a way to study nuclear morphology, which is an indicator of cell health. For instance, unhealthy and dying cells exhibit condensed nuclei, whereas healthy cells have nuclei that are somewhat oval shaped.

In addition to providing fluorescent labels, small-molecule fluorophores also have applications in the recognition of biological analytes. These fluorophores are often called detectors reporters, or sensors. Many fluorescent small molecules that monitor pH change, enzymatic activities, and analytes (biological small molecules like peroxide, metal ions, and gases like NO) in cell culture have been published. Many of these detectors contain fluorophores like fluorescein, rhodamine, and BODIPY. In my PhD work, I attached metal ion chelators to fluorescein and used them to detect zinc in live cells and brain tissue. Some academic laboratories actively designing biocompatible small-moleule fluorescent sensors include those of Professor Stephen J. Lippard at MIT, Professor Christoph Fahrni at GeorgiaTech, Professor Christopher J. Chang at University of California, Berkeley, Professor Kazunori Koide at University of Pittsburgh, and the Center for Molecular Imaging Research at Harvard Medical School. There are many other laboratories across the globe working on fluorescent small-molecule fluorescent sensors. If you look on a web-based science literature search engine like PubMed (and I urge you to do so), you'll likely find dozens of reviews addressing this general topic. You'll also learn that fluorescent sensors are by no means limited to biological imaging—there are plenty of other exciting uses for these molecules, but I need to save such stories for another time!

## SITE-SPECIFIC LABELING OF PROTEINS WITH SMALL MOLECULES FOR IMAGING

Hopefully, I have convinced you that small fluorescent molecules are really useful for biological imaging. Nevertheless, they do have some limitations. Often, it can be difficult to predict how a small molecule will behave in cell culture. Will it permeate the cell membrane? If so, where will it go? If not, can we make modifications synthetically to provide permeability (as with esterification of fluorescein)? Some problems with many small-molecule fluorophores are that they spontaneously localize in cells, redistribute within cells, or even leak out of cells over time. These behaviors can be problematic depending on the experiment and imaging protocol. Imagine wanting to conduct an all-day imaging experiment using a dye that leaks out of cells after only a few hours! Thankfully, a number of chemists have devised ways to solve this problem.

Several research groups have worked to tackle problems associated with fluorophore localization and movement by developing methods to site specifically and covalently link fluorescent probes to proteins within or on the exterior of cells (Figure 8.3). These approaches generally rely on attaching a small-molecule probe with a reactive group to a specific peptide/protein that is expressed in a desired cellular location (like with the GFP–protein conjugates I mentioned, molecular biology techniques are used in this work). Because a covalent bond is formed between the peptide/protein and the label, the dye cannot spontaneously and uncontrollably disperse in the cell. The labeling is called "site specific" because, assuming no side reactions, the chemical reaction only occurs at the protein/peptide site expressed.

Professor Roger Tsien's group pioneered a method that uses a small molecule named FLAsH (Figure 8.3). FLAsH is a fluoroescent biarsenical ("FL" is for fluorescein and "As" is for arsenic) compound. Arsenic has a high affinity for thiol residues (a thiol group is –SH) and FLAsH reacts with a tetracysteine peptide motif Cys-Cys-X-X-Cys-Cys to provide covalent linkages (Cys stands for the amino acid cysteine, X stands for any other amino acid and preferably proline-glycine, and the hyphens mean that these amino acids are linked together by peptide bonds). This six amino acid sequence, called a "tag," can be fused to a protein and expressed at specific locations within cells. FLAsH is cell permeable and can bind to the tetracysteine peptide motif to provide a covalently attached fluorescent label.

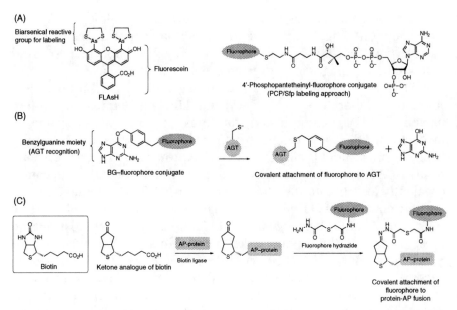

**Figure 8.3.** Examples of small molecules employed in site-specific labeling methods. (A) Structures of FLAsH and a 4′-phosphopantetheinyl–fluorophore conjugate. (B) Structure of a BG–fluorophore conjugate and a scheme for the covalent attachment of the fluorophore to AGT. (C) Structures of biotin and the ketone analogue of biotin. Attachment of the ketone analogue to an AP–protein fusion and subsequent reaction with a fluorophore hydrazide for covalent fluorophore attachment.

Since its initial use in the late 1990s, FLAsH analogues have been prepared with different excitation and emission wavelengths and/or with additional reactive groups for other applications. Like most methods, however, FLAsH has some drawbacks. The reactivity of the biarsenical is not completely specific for the Cys-Cys-X-X-Cys-Cys motif. Arsenic is thiophilic and, as a result, FLAsH can react nonspecifically with other cysteine residues in the cell. FLAsH can therefore potentially label a myriad of proteins within the cell, which results in background fluorescence and might complicate data anlaysis.

Several strategies have therefore been developed to overcome nonspecific labeling in cells that results from side reactions of the fluorophore with other biomolecules. Like the FLAsH approach, these methods employ tags that are proteins or small peptides, and the tags are fused to a protein and expressed at desired cellular localizations. Then, a fluorophore with a reactive moiety capable of covalent attachment to the tag is introduced and site-specific labeling is achieved. Two important factors in the development of these methods are (1) the size of the tag and (2) the selectivity of the attachment. It's generally accepted that a smaller tag is superior to a larger one because the small size is less likely to perturb the biological system significantly. A highly selective reaction that supplies the covalent linkage between the tag and fluorescent label is important to avoid nonspecific labeling and background fluorescence.

One approach that utilizes a protein tag comes from Professor Kai Johnsson's laboratory in Switzerland. Humans express a protein named $O^6$-alkylguanine-DNA transferase (AGT). This protein is involved in DNA repair and it irreversibly transfers the alkyl group from its substrate, $O^6$-alkylguanine-DNA, to one of its own cysteine residues. As a result, the alkyl group originally on DNA becomes covalently attached to a cysteine residue in AGT. The Johnsson laboratory found that AGT will accept many synthetic small molecules as substrates if an $O^6$-benzylguanine (BG) moiety is attached, including those containing fluorophores like fluorescein and rhodamine (Figure 8.3). When a fluorophore–BG conjugate is employed, the fluorophore becomes covalently attached to AGT. AGT can be overexpressed in cultured cells and, by using molecular biology to incorporate targeting sequences, its expression can be directed to the cytoplasm, organelles, and the cell membrane. Subsequent treatment of cultured cells overexpressing AGT with fluorophore–BG conjugates results in covalent attachment of the fluorophore and site-specific labeling. Recently, this method has been used to site specifically localize Zn(II) and Ca(II) sensors within cells. One possible disadvantage of this method is the size of the AGT tag. AGT is >200 amino acids, much larger than the tetracysteine motif used for FLAsH labeling. Also, because mammalian cells express AGT, endogenous AGT might complicate things and cause background fluorescence.

Another approach comes from Professor Christopher T. Walsh's laboratory at Harvard Medical School. The Walsh laboratory studies the biosynthesis of natural products (natural products are small molecules created by nature). What does natural product biosynthesis have to do with cellular imaging? Like the AGT method where a DNA repair protein is used, it turns out that some of the proteins involved in natural product assembly are useful for labeling the cell surface. Peptide carrier proteins (PCPs) are 80- to 120-amino acid domains of nonribosomal peptide synthetases (NRPSs). NRPSs are protein megacomplexes used by many microbial species, like *Pseudomonas* and *Streptomyces*, to biosynthesize natural products from common amino acid precursors. An enzyme called a phosphopantetheinyl transferase will covalently attach the 4′-phosphopantetheinyl moiety of coenzyme A to a specific serine residue in the PCP domain. One phosphopantetheinyl transferase, Sfp from a microbe called *Bacillus subtilis*, can transfer 4'-phosphopantetheinyl–fluorophore conjugates to PCP domains. As you can probably guess, this promiscuity allows Sfp and 4'-phosphopantetheinyl–fluorophore conjugates (Figure 8.3) to be used for the labeling of protein–PCP fusions. As I mentioned above, PCP domains contain 80–120 amino acids, making them smaller than the AGT label but still relatively big. In order to overcome this potential limitation, the Walsh group identified a short, 11-residue peptide tag that is a substrate for Sfp. As a result, only 11 additional amino acids must be incorporated into a protein for labeling. In terms of cellular imaging, this methodology is limited to the cell surface because Sfp and 4'-phosphopantetheinyl–fluorophore conjugates cannot permeate the cell. Nevertheless, there are many applications for labeling proteins at the cell surface. This PCP/Sfp method was used to fluorescently label the transferrin receptor.

I'll tell you about one more labeling method employed in cell imaging studies. The approach, developed by Professor Alice Y. Ting at MIT, allows for labeling of the cell surface and uses the enzyme biotin ligase (BirA) from *Escherichia coli*. BirA catalyzes the attachment of biotin to a lysine residue incorporated in a 15-residue peptide named "acceptor peptide" (AP). The Ting laboratory showed that BirA also accepts a synthetic ketone analogue of biotin as a substrate (Figure 8.3). Ketone functional groups are absent from biomolecules like lipids, proteins, and sugars, and they react with hydrazides. As a result, once the ketone analogue of biotin is covalently linked to AP expressed on the cell surface, a fluorophore-hydrazide can be added to the cell culture and it will only react with the ketone analogue to provide site-specific attachment of the fluorophore. Like FLAsH and the PCP/Sfp approach, this method employs a relatively small peptide tag. In contrast to methods where the fluorophore label is attached in one step, the biotin ligase approach requires two reactions to incorporate the fluorescent label: (1) BirA-catalyzed attachment of the ketone analogue of biotin and (2) reaction of the ketone with a fluorophore-hydrazide probe. I should note that the natural substrate for BirA, biotin, provides another means for site-specific fluorophore attachment. Streptavidin is a protein that binds to biotin with very high affinity, and it can be conjugated to many things including fluorophores and quantum dots (quantum dots are semiconducting nanocrystals that provide bright fluorescence emission). Following BirA-catalyzed attachment of biotin to the AP, streptavidin conjugates can be added to incorporate a label via the biotin–streptavidin interaction. The Ting laboratory used this method in fluorescence imaging studies of neurons where the AMPA (alpha-amino-3-hydroxy-5-methyl-4-isoxazoleproprionic acid) receptor was labeled with quantum dots.

I hope that these four examples provide some idea of the diverse and creative approaches chemists have developed to site specifically and covalently attach fluorescent labels for cellular imaging. It will be exciting to see how these methodologies are employed in future studies of biological phenomena like cell signaling and neurophysiology, and also to learn how else chemists decide to tackle the challenge of controlling probe localization. Also, because I focused on small-molecule fluorophores, I should note that these methods can be used to attach other "things" like quantum dots, affinity probes, or other small molecules that might be of interest.

## MRI

Now let's leave fluorescence behind for a bit. Cellular imaging is not limited to fluorescence-based approaches, and chemists also contribute to alternative imaging techniques like MRI. Did you think about how chemistry might influence MRI after your bike accident and subsequent trip to the doctor last summer? If not, think again!

As you experienced with your knee injury, MRI is frequently used in diagnostic medicine. The technique is nonradioactive and noninvasive, which makes it ideal for the clinic. It relies on differences or "contrast" in nuclear magnetic resonance (NMR) that occur at different locations in a sample. You'll learn about NMR in organic chemistry this year, and MRI is based on these principles. In brief, different protons in the body (generally the protons of water molecules) will have different magnetic fields associated with them depending on their local environment, and these differences will produce variations in the MRI signal. In the brain, for instance, MRI can differentiate white from gray matter. Sometimes, however, the resolution of different anatomical features or cells is not possible with MRI alone. In these cases, "contrast agents" must be administered in order to obtain an informative MRI. Contrast agents are often synthetic small molecules or iron oxide nanoparticles, and they must alter the NMR properties of water molecules. Then, an MRI scan will detect the variations in the NMR properties of the water molecules in the organ/tissue/body that result from the administration of the contrast agent. Because contrast agents are administered to patients, they must be biologically compatible (low toxicity, water solubility, quickly excreted, etc.).

Many small-molecule MRI contrast agents contain lanthanide ions. Lanthanides are often called "rare earth" elements, and they comprise elements 57–71 of the periodic table. Lanthanides, and gadolinium(III) in particular, are useful for MRI because their electronic and magnetic properties are well suited for the technique. The Gd(III) ion will be bound by a chelator (a chelator is a metal-coordinating molecule) that confers any number of features—water solubility, intracellular localization, magnetic properties, and so on. One very important feature of many Gd(III)-based contrast agents is that a water molecule is coordinated to the Gd(III) center. Once *in vivo*, this water molecule can exchange with other water molecules in the surrounding environment. This swapping changes the NMR of water molecules in the location of interest and thereby generates the MRI signal. Some academic laboratories that synthesize and study lanthanide-containing MRI contrast agents include those of Professor Thomas Meade at Northwestern University and Professor Kenneth Raymond at the University of California, Berkeley.

Polyaminocarboxylate ligands are often employed for Gd(III)-based MRI contrast agents, including some that are in clinical use today (Figure 8.4). These ligands are good choices because they provide very stable complexes (the multiple carboxylates bind to the Gd(III) center) with low toxicity. One contrast agent with a simple structure is Gd-diethylenetriaminepentaacetic acid (DOTA). Modifications to Gd-DOTA have provided MRI contrast agents with additional properties or functionalities. For instance, the Meade Laboratory covalently linked a steroid, RU-486, to Gd-DOTA. The steroid binds selectivity to the progesterone (a steroid) receptor. Efforts have also been made to attach fluorophores to Gd-DOTA and other MRI contrast agents. These types of conjugates allow for the imaging of one sample using fluorescence and MRI techniques, which could be of benefit in the identification and surgical removal of tumors.

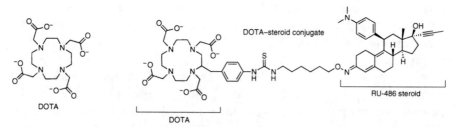

**Figure 8.4.** *Structures of the DOTA ligand and a DOTA–steroid conjugate. DOTA binds Gd(III) with the four tertiary amine nitrogen atoms and the four carboxylate oxygen atoms and provides an open site for a water molecule to coordinate.*

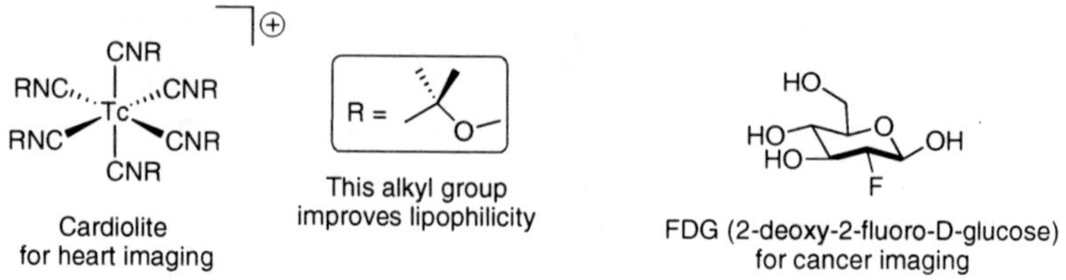

CNR
$\oplus$

RNC,,,,  CNR
RNC  Tc  CNR
CNR

Cardiolite
for heart imaging

R = 

This alkyl group
improves lipophilicity

HO
HO  O
HO  OH
F

FDG (2-deoxy-2-fluoro-D-glucose)
for cancer imaging

**Figure 8.5.** *Small molecules used in nuclear medicine.*

In addition to Gd(III), iron oxide nanoparticles and manganese-containing small molecules are useful for *in vivo* MRI. There is also substantial interest in designing MRI-based sensors. Like fluorescence-based sensors, MRI sensors can be used to monitor/detect pH change, metal ions (Ca, Zn, and Cu), enzymatic activity, and other biological phenomena.

## NUCLEAR MEDICINE AND USING RADIOACTIVITY FOR IMAGING

Lastly, let's consider some small molecules that are used in nuclear medicine and also mention some particular challenges chemists face in this field. Nuclear medicine involves the use of radioisotopes to detect and treat disease. In terms of detection, radioactive "tracers" can be employed to provide images of the body, organs, and diseased tissues. I want to tell you a bit about radioactive tracers and their synthesis. Chemists have made, and continue to make, great contributions in this area.

Have you heard of Cardiolite (Figure 8.5)? Maybe your introductory chemistry professor mentioned it as a coordination compound with applications in medicine? It's a radioactive heart-imaging agent and contains the radioisotope $^{99m}$Tc (the "m" is for "metastable") and provides a wonderful example of how fundamental inorganic chemistry can be applied to biological imaging. Professor Alan Davison, an emeritus chemistry professor at MIT, discovered this compound. The technetium center of Cardiolite is coordinated in an octahedral fashion by six cyanide (CN⁻) ligands. The cyanide ligands each contain an alkyl group to make the molecule adequately lipohilic. The $^{99m}$Tc isotope is a gamma emitter and has a half-life of only about 6 hours. The half-life is the amount of time it takes for $^{99m}$Tc, or any radioactive element, to lose half of its initial radioactivity. These properties are important from the standpoint of therapeutics because gamma rays are easily detectable by X-ray imaging and, due to the short half-life, the radioactivity is not prolonged in the body. On the other hand, these features create some significant challenges for a chemist seeking to prepare new technetium-based diagnostics! First, doing synthesis with toxic and radioactive substances requires certain precautions to avoid exposure. In addition, any chemical synthesis needs to be completed very quickly because the half-life of $^{99m}$Tc is very short. It would be a shame if the tracer decayed prior to its use in a patient! What can an experimental chemist do to overcome these problems? If we look at the periodic table, we see that technetium is a group 7 element and is located between manganese (Mn) and rhenium (Re). Rhenium is much more stable than Tc and is a good Tc analogue. Synthetic chemists can first prepare models of new Tc-based imaging agents using Re. After the synthesis and purification are optimized with Re, the synthesis can be repeated with $^{99m}$Tc to generate the novel tracer.

*In vivo* imaging techniques employed in nuclear medicine aren't limited to the use of technetium. Another important nucleus for diagnostics is $^{18}$F. This isotope is used as a tracer for positron emission tomography (PET). Like $^{99m}$Tc, $^{18}$F is a gamma emitter. Its half-life is very short, only about 2 hours. Therefore, the isotope must be generated using a cyclotron that is very close to the medical imaging facility. One important diagnostic that contains $^{18}$F is 2-fluoro-2-deoxy-D-glucose (FDG, Figure 8.5), and it is used for detecting cancerous lesions. Cancer cells require large amounts of glucose, and so they will take up and metabolize more FDG than surrounding normal cells. This differential uptake will provide the contrast observed in a PET image of a tumor.

What is required for development of new $^{18}$F-containing radiotracers and how can chemists contribute? Because of its very short half-life, $^{18}$F must be introduced into radiotracer precursors very efficiently and at a very late stage (if not the final stage!) of the synthesis. One way to incorporate $^{18}$F into an organic molecule is to form a C—F bond, but this chemistry is pretty tricky and relatively unexplored. Unlike the myriad of known reactions that mediate C—C, C—N, and C—O bond formation (you'll learn about some in organic chemistry this year), few methods for creating C—F bonds exist and some aren't very efficient. Therefore, fundamental studies of C—F bond formation are important for the development of $^{18}$F-containing molecules for nuclear medicine. Thankfully, several laboratories, including the groups of Professor Melanie Sanford at the University of

Michigan and Professor Tobias Ritter at Harvard University, are addressing this problem and elucidating exciting new chemistry along the way. For instance, both groups recently reported that palladium complexes can be used to mediate C—F bond formation.

From these examples of radiotracers, I hope you now appreciate that fundamental organic and inorganic chemistry are invaluable to nuclear medicine and diagnostics. Chemistry that might seem far removed from imaging—for instance, formation of C—F bonds—at first glance can make tremendous impact on the field.

Well, Angela, I think that's about it for now. I didn't anticipate that this letter would get so long and it's getting late. The coffee shop I'm sitting in is about to close! As I mentioned earlier, I decided to pursue research outside of the imaging field in my postdoc. But I hope I've convinced you that the field of cellular/molecular imaging is an exciting one and that there are plenty of ways for chemists to contribute. Please let me know if you want any suggestions for references or reading. I hope to see you at our family reunion next year, if not before! If you travel to Boston sometime, please let me know. I'd be happy to show you around the city and, of course, the laboratory. Best wishes for the rest of your semester. Work hard and have fun!

Take care,

Liz

## FURTHER READING

Bottrill, M.; Kwok, L.; Long, N. J. Lanthanides in magnetic resonance imaging. *Chemical Society Reviews* 2006, *35*, 557–571.

Chen, I.; Ting, A. Y. Site-specific labeling of proteins with small molecules in live cells. *Current Opinion in Chemical Biology* 2005, *16*, 35–40.

Haugland, R. P. *Handbook of Fluorescent Probes and Research Products*, 9th Edition. Molecular Probes, Inc., Eugene, OR 2002. Available at http://www.probes.com.

Johnsson, N.; Johnsson, K. Chemical tools for biomolecular imaging. *ACS Chemical Biology* 2007, *2*, 31–38.

Nobel Prize Lectures for the 2008 Nobel Prize in Chemistry to Osamu Shimomura, Martin Chalfie, and Roger Y. Tsien. Available at http://nobelprize.org/nobel_prizes/chemistry/laureates/2008/index.html.

# 9

# Bioinorganic Chemistry: Show Your Mettle by Meddling with Metals

**Kara L. Bren**

*University of Rochester*

Kara L. Bren is a professor of chemistry at the University of Rochester in Rochester, New York. She earned her BA in chemistry at Carleton College in Northfield, Minnesota, where she did research with Professor Lynn Buffington on the nuclear magnetic resonance (NMR) of carbohydrates. She earned her PhD in the lab of Professor Harry B. Gray at the California Institute of Technology in Pasadena, California, studying ligand binding properties of engineered heme proteins. While earning her graduate degree, she spent time in the lab of Ivano Bertini at the University of Florence, where she learned about the NMR of paramagnetic biomolecules. She continued her training as a National Institutes of Health (NIH) postdoctoral fellow in the lab of Professor Gerd N. La Mar at the University of California, Davis, where she performed NMR studies on paramagnetic iron–sulfur proteins. At Rochester, the Bren group is utilizing a range of spectroscopic techniques to study the molecular and electronic structure and folding of heme proteins in the cytochrome *c* family.

*Letters to a Young Chemist*, First Edition. Edited by Abhik Ghosh.
© 2011 John Wiley & Sons, Inc. Published 2011 by John Wiley & Sons, Inc.

Dear Angela,

I trust that you are readjusting to life back at University of California, San Diego (UCSD) after your stay in Norway. I am so glad that Abhik realized that you would be traveling through Oslo while I was visiting my collaborator there and arranged for us to meet. Norway certainly was beautiful, and I look forward to future visits. When I was a student considering a career as a chemist in academia, I never imagined that a perk of my choice would be the many opportunities for travel. Between going to conferences and visiting collaborators, this profession gives you an unexpected chance to see the world.

Since our time together was short, we didn't get nearly as much of a chance to talk as we would have liked. So, as promised, I am following up now to tell you more about the field of bioinorganic chemistry and the kinds of things scientists working in this field study. Simply speaking, bioinorganic chemistry is the study of the chemistry of the elements in biological systems. The field includes the study of metallic and nonmetallic elements in biology, although much of the focus is on metals, as indeed they constitute the bulk of the periodic table; as a result, I will often use the term "metals" in my discussion of the field. I understand that you, like many undergraduate students, have not yet had a bioinorganic chemistry course and that you had heard just a little about bioinorganic chemistry. You mentioned that you had learned in your introductory chemistry course that iron plays a vital role in carrying oxygen in your blood. Outside of class, you'd heard news stories about the dangers of toxic lead pigments and the controversy over mercury in vaccines, which paint a more sinister picture of the roles of metals in biology. If only the positive roles were also deemed newsworthy! Nevertheless, these examples all illustrate different facets of bioinorganic chemistry, although there is much more.

Given the breadth of this field, I have to force myself to pick out just a few areas to tell you more about, because if I wrote about everything in the field, this would be an encyclopedia rather than a letter! First, I thought I'd write about three topics that would give you some general information on the field: (1) an overview of why and how biological systems make use of metals, (2) examples of roles played by metal-containing proteins (metalloproteins) in the biosphere, and (3) methods for detecting and studying metals in biological systems. In addition, I wanted to tell you about some of the areas you mentioned you'd heard a bit about in other courses and while surfing the Web, such as (1) the role of metal ions in Alzheimer's disease (AD) and (2) metal-based drugs. This should give you both a broad introduction to the kinds of molecules we study and the techniques we use, and some specific examples relevant to human health of studies of both naturally present and introduced metals in biological systems.

## INORGANIC LIVES!

If you decide to pursue a career as a bioinorganic chemist, one thing you will encounter is the confusion the term "bioinorganic" brings about, especially among nonscientists but even among other scientists. "Did you mean to say that you study bio-*organic* chemistry?" people often ask me, puzzled since the term "organic" is commonly associated with living things, whereas "inorganic" is associated with nonliving things. To the contrary, inorganic chemistry and life are not only completely compatible, but life *requires* elements and molecules that fall outside the purview of organic chemistry. To illustrate this point, I've included a periodic table indicating the elements essential for life, and you will see that there are many elements used aside from carbon, hydrogen, nitrogen, oxygen, sulfur, and phosphorus, which are those on which most biochemistry texts focus (Figure 9.1).

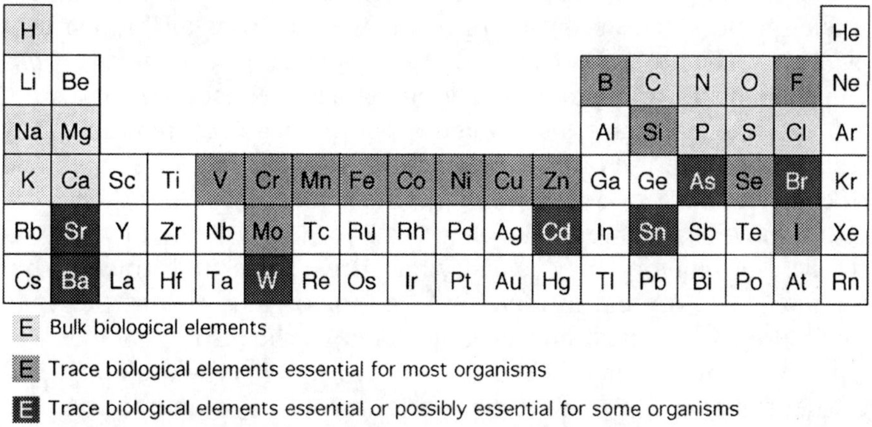

| H | | | | | | | | | | | | | | | | | He |
|---|---|---|---|---|---|---|---|---|---|---|---|---|---|---|---|---|---|
| Li | Be | | | | | | | | | | | B | C | N | O | F | Ne |
| Na | Mg | | | | | | | | | | | Al | Si | P | S | Cl | Ar |
| K | Ca | Sc | Ti | V | Cr | Mn | Fe | Co | Ni | Cu | Zn | Ga | Ge | As | Se | Br | Kr |
| Rb | Sr | Y | Zr | Nb | Mo | Tc | Ru | Rh | Pd | Ag | Cd | In | Sn | Sb | Te | I | Xe |
| Cs | Ba | La | Hf | Ta | W | Re | Os | Ir | Pt | Au | Hg | Tl | Pb | Bi | Po | At | Rn |

E  Bulk biological elements

E  Trace biological elements essential for most organisms

E  Trace biological elements essential or possibly essential for some organisms

**Figure 9.1.** *Periodic table indicating elements utilized by living organisms. The lanthanides and actinides are omitted because none have been found to be essential for life. Many additional elements are utilized in probes and drugs. Adapted from Bertini et al. (2007).*

You probably encountered the importance of inorganic chemistry for your own health at a young age even if you don't realize it. Indeed, you may have been told as a child that eating spinach gives you strong muscles and that drinking milk gives you strong bones. If your parents were well versed in nutrition, they might have added that this is because spinach is a source of iron, which plays many roles related to muscle function, and milk is a source of calcium, a vital component of the biomineral we know as bone. There are other elements required for life that don't get as much attention at the dinner table. For example, people don't usually think of cobalt as a nutrient, but it is an essential component of vitamin $B_{12}$, which is found in meat, eggs, and dairy products, and is required to form healthy blood and nerve cells. Living things require other metals that sound exotic, such as manganese, selenium, and molybdenum. In addition to essential metals, organisms will sometimes take up elements that are not essential for life but that will have a physiological effect. For example, a cancer patient may take a platinum-based chemotherapeutic drug, or a child may ingest lead-harboring paint chips while playing in the sandbox. The study of how these nonessential metals affect organisms, both positively and negatively, also falls within the field of bioinorganic chemistry. It was my experience that once I became aware that metals are important in biology, I started noticing them everywhere. It is like when you first learn a new vocabulary word, and then suddenly realize you heard it on the news, read it in a book, and so on.

Bioinorganic chemistry is often considered to be a specialized field. "Specialized," however, should not be confused with "narrow" because bioinorganic chemists must utilize knowledge from all branches of chemistry (Figure 9.2). They, of course, must know about the *inorganic* chemistry of metal ions (and nonmetallic elements), but also of the *organic* molecules and moieties with which they interact. They frequently make use of *physical* methods to make measurements, and of course either work with *biological* molecules and/or put their work into a biological context. The range of knowledge needed to work in the field of bioinorganic chemistry may drive some to distraction, but for me, it was a great attraction. As a bioinorganic chemist, I get to make use of the latest advances in different chemistry subfields as I work with my students in the lab, which I find very stimulating. This also encourages me to develop a diverse toolbox of methods to use to attack interesting problems, which is a liberating and productive way to do science. It can be a challenge, however, to keep up with the latest advances in a broad and interdisciplinary field, which is

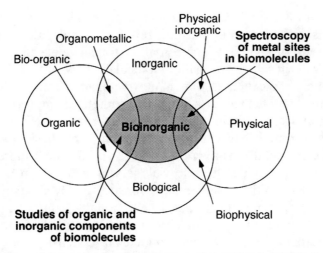

**Figure 9.2.** *Venn diagram showing different areas of chemistry in relation to bioinorganic chemistry (shaded area). The interdisciplinary areas most closely related to bioinorganic chemistry are in bold. This diagram depicts just one possible way to envision the interrelatedness of these different fields within chemistry.*

why I like to say that while all life requires metals, bioinorganic chemists also require mettle! On the other hand, the breadth of bioinorganic chemistry as a field allows scientists to find a component of the field on which to focus that fits their strengths and interests, be it synthesis, spectroscopy, theory, biochemistry, or even cell biology. You could say that there is something for everyone.

Well, I hope I've made the point that inorganic doesn't have to mean nonliving, so the next question that comes to a chemist's mind is *why* so many elements, including metals, are essential for life. Biomolecules have diverse structures and functions, and chemical versatility is essential for life. Considering this, it is easy to see why life is based on carbon, as it has highly diverse chemistry, being able to bond in linear (as in acetylene), trigonal planar (as in ethylene), or tetrahedral (as in hydrocarbons) fashions, form compounds with most other elements on the periodic table, and form stable small molecules, rings, and high-molecular-weight polymers. No other element has all of these attributes. Despite the versatility of carbon, there are a number of things that organic molecules cannot easily do, at least without some assistance, that are easy for inorganic molecules. For one, carbon cannot bond to more than four different groups, whereas metal ions may bind as few as 1 and as many as 8 or 10 groups (often called ligands). In addition, carbon forms strong, directional bonds, and while this property is beneficial for stability, it makes bond making and bond breaking challenging. In contrast, metal ions are capable of forming relatively weak but stable bonds with a variety of ligands, which can allow for facile exchange to occur. Another way in which inorganic molecules show differing reactivity from organic molecules is that many metal ions can readily accept and donate one electron at a time, and sometimes more than one in sequence, and can remain stable (relatively unreactive). This kind of activity is essential in processes such as photosynthesis and respiration, in which metal ions play essential roles (although there also are some organic molecules that do well serving as electron donors and acceptors). The ability of some metal ions to readily change oxidation state by two or more units makes them valuable components of biomolecules that catalyze multielectron transformations. Finally, metal ions are capable of achieving a high charge density, meaning that they can carry positive charge in a relatively small volume. This makes metal ions good Lewis acids, or electron pair acceptors, which allows them to activate other molecules as needed to carry out reactions.

As an example of the special properties of metals, consider the function of the oxygen-carrying protein hemoglobin, which has an iron that binds either five (in the deoxygenated form) or six (in the oxygenated form) groups. It is capable of binding oxygen gas in a reversible fashion, quickly binding it in oxygen-rich tissues and releasing it in oxygen-poor tissues. This is all the more remarkable because oxygen is a strong oxidizing agent and generally does not react reversibly with organic molecules (in fact, we refer to this reaction as combustion!) When oxygen binds, the hemoglobin iron changes its effective oxidation state from +2 to +3 and then back to +2 when oxygen is released. The properties of the metal ion are essential for hemoglobin's function.

## METALS, METALS, METALS, ALL THE WAY DOWN

Given the reactivities displayed by metals, it is not surprising that many biological catalysts, known as enzymes, make use of metal ions (referred to as metal cofactors). Enzymes can be nucleic acids, but most are proteins, and I will be referring to protein enzymes in this letter. Enzymes enhance reaction rates by lowering energy barriers between reactants and products, and indeed the ability to catalyze reactions is considered to be a requirement for all life-forms. Organisms must catalyze reactions in order to break down nutrients, construct macromolecules, pump ions across membranes, repair damaged genes, or undergo photosynthesis. With their abilities to bind and release ligands, engage in electron transfers, and act as Lewis acids, metal ions play key roles in all of these processes, and indeed, approximately 30–50% of proteins bind metal cofactors.

One way to illustrate this point is to consider biogeochemical cycles of the elements. This is a topic on which I saw Professor Ed Stiefel give a beautiful talk early in my career, and it made an impression on me because it illustrated the interconnectedness of the geosphere and the biosphere, and highlighted the critical roles played by metalloenzymes in these cycles. What is meant by biogeochemical cycles are sets of interconnected catalyzed reactions that transform elements essential for life (including carbon, hydrogen, nitrogen, oxygen, sulfur, and phosphorus) into different forms. The cycle you have probably heard the most about is the nitrogen cycle, with its most famous step being nitrogen fixation. In nitrogen fixation, the enzyme nitrogenase catalyzes a remarkable reaction in which the inert dinitrogen ($N_2$) molecule is reduced by six electrons to produce two molecules of ammonia ($NH_3$), a form of nitrogen useful for most life-forms. Fixation is only one step of the nitrogen cycle, which is illustrated in Figure 9.3, with the types of metal ions involved in

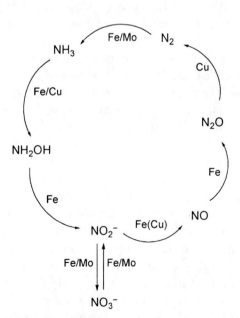

**Figure 9.3.** The nitrogen cycle, highlighting the metal cofactors involved in each step. Adapted from Morel, F. M. M.; Price, N. M. The Biochemical Cycles of Trace Metals in the Oceans. Science 2003, 300, 944–947.

catalyzing each step indicated; you will note that each step requires metal cofactors. Determining the structure, properties, and mechanism of action of the enzymes required for this cycle is an example of an area of current intense investigation in bioinorganic chemistry. Considering the ways in which metalloenzymes link one reaction to another in the nitrogen cycle, I am reminded of a story about a scientist (the name of whom changes in different versions of the story) who, after a talk on the nature of the universe, was approached by an elderly woman who told him that the earth does not float in space but is carried on the back of a turtle. When the scientist, thinking he'd quickly change her mind, asked her what holds that turtle up, her reply was "it's turtles, turtles, turtles, all the way down!" In the nitrogen cycle, it is metals, metals, metals, all the way down, with each metalloprotein-catalyzed reaction depending on another, but none of them being a true starting point.

## DO YOU SEE WHAT I SEE?

Some of the early studies in bioinorganic chemistry were inspired by what scientists literally saw with their own eyes. The intensity of the blue color characterizing blue copper proteins led inorganic chemists to realize that there was something special about the environment of the copper ion in these proteins, and the discovery of heme proteins called cytochromes was facilitated by their being the most intensely colored substances isolated from cells ("cyto" and "chrome" mean "cell" and "color," respectively). The appearance of color is a result of the selective absorption of particular wavelengths of visible light by a substance, and this selective absorption of light according to color (which corresponds to frequency) also describes a type of spectroscopy. Spectroscopy is the study of the interaction of electromagnetic radiation with matter and comes in many different flavors. Absorption of visible light as well as ultraviolet (UV) and near infrared is measured by electronic spectroscopy, but this represents a small portion of the electromagnetic spectrum (Figure 9.4). Spectroscopic methods utilize different energies of radiation and thus probe varied properties of biomolecules. An example of a spectroscopic method that utilizes low-energy radiation (in the radiofrequency range) is NMR, which probes transitions of individual nuclear spins. This method is useful for getting structural information because it reports on the environment of each nucleus. To get information

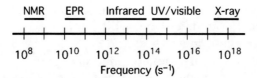

**Figure 9.4.** *Frequencies (in s⁻¹) of radiation for some spectroscopic methods employed by bioinorganic chemists. High frequency corresponds with high energy.*

about electron spins, one needs to utilize slightly higher energy microwave radiation, which is used in electron paramagnetic resonance (EPR) spectroscopy. On the other end of the spectrum are high-energy spectroscopic methods such as X-ray absorption spectroscopy (XAS), which utilizes high-energy radiation to excite tightly held electrons and provides information about the local structure around a specific metal. Bioinorganic chemists have from the founding of the field played roles in the development and application of new spectroscopic methods. The interesting and unusual environments in which metals find themselves in biological molecules cause them to take on properties that call out for spectroscopic study. In addition, methods that directly probe the metal itself provide a way to "isolate" the metal site from the rest of a large biomolecule, or even from other components within a live cell. This is a way to obtain highly detailed information about a metal site even when it is just one tiny component of a very complex system.

The information you can get from spectroscopy of course depends on the particular method you use. You may be able to determine molecular structure, metal site symmetry, molecular motions, bond vibration energies, orbital energies, distances between specific nuclei, and energies of interactions between electrons, just to name a few. Spectroscopic methods constitute a powerful toolkit! An important thing to keep in mind is, as with all measurements, what you see depends on how you look at something. Think of the fable of a group of people examining an elephant by touch only, where each touches a different part of the elephant (trunk, tusk, and leg), and thus each comes to a completely different conclusion about the properties of the animal. Similarly, using one spectroscopic method may give you a different apparent answer than another, which is why bioinorganic chemists often utilize many different methods. In some cases, though, we have few methods we can use at all. Metals that have "closed shells" (all valence orbitals filled) are difficult to probe using spectroscopy. Unfortunately, this includes a number of metals that are important biologically, such as calcium, magnesium, and zinc. New methods, which I'll comment on a bit later, are being developed to aid the study of these metals.

For most of its history, bioinorganic chemistry experiments have been performed on purified samples of metallobiomolecule concentrated in a buffered aqueous solution (although some experiments are done on frozen samples or solids). In addition, the concentration of biomolecule and the medium containing it used in an experiment are generally selected according to the requirements of the experiment rather than according to what the molecule's native environment may be. This isn't an unreasonable approach because biological macromolecules are already quite complicated, and to obtain detailed information on their structure and function, it is helpful or even essential that you do not have other substances present that could interfere with making measurements and/or interpreting data. And in any case, we rarely, if ever, know the true environment a biomolecule experiences as it functions within a cell. We do know, though, that the cellular environment is very different from a simple buffered aqueous solution. The interior of a cell is extremely "messy," packed with various compartments, organelles, and a whole slew of macromolecules and metabolites including species that might be reactive (such as oxidants or reductants) and that may perturb the structure and activity of biomolecules (such as ligands and membranes). Bioinorganic chemists have been thinking more than they used to about inorganic chemistry within the cell, which includes the properties and behavior of metallobiomolecules in a cellular environment as well as the question of how cells regulate levels of metal ions within them. The fact that there is a limited "healthy" concentration range below which there is deficiency, and above which there is toxicity, for all metal ions both on the organismal and cellular levels makes this second question interesting in the context of inorganic chemistry and medicine.

Studying metal ions and metallobiomolecules within cells requires the development of methods that allow specific, sensitive, and rapid metal ion detection within the complex cellular environment. Some studies in this area have focused on answering the deceptively simple question "What is the concentration of a metal in the cell?" Such studies have utilized classic inorganic chemistry techniques cleverly applied to a cellular environment. Indications are that the levels of free (not bound to other molecules) metal ions are very low, suggesting tight control of metal ion levels in the cell. Interfacing with the expertise of biological scientists has helped address the follow-up question "How does the cell control the levels of metal ions?" By combining methods from the fields of genetics, inorganic chemistry, and structural biology, great headway has been made in understanding how metal ions are taken up and excreted by cells and how cells achieve tight control of metal ion levels. One discovery is that cells utilize proteins called metallochaperones to bind and transport metals within the cell. It also has been learned that proteins act as metal ion sensors in the cell, binding a specific metal ion and undergoing a change in conformation that triggers other processes in the cell related to controlling the levels of that metal ion. This work has been done on metal ions such as zinc, manganese, iron, and copper, and also on the iron-containing heme group.

A distinct but related approach to investigating inorganic chemistry in the cell involves the development of cell-permeable chemical probes for metal ions. This tactic is particularly important for the "spectroscopically silent" metal ions. These molecules also are called sensors because they provide a signal about the presence of a metal ion to the experimenter. The design of metal ion sensors presents a significant challenge to the inorganic chemist because of their many requirements. I will use zinc sensing as an example because it has attracted a lot of attention recently. The sensor must first display a substantial change in luminescence upon zinc binding, preferably with an increase in the presence of the ion (it is difficult to interpret the loss or absence of signal). Luminescence is the preferred signal because it can be detected even on the level of individual molecules and in the presence of other molecules and structures within the cell. The sensor must be selective for zinc over all other metal ions and other components within the cell, including those present at much higher concentrations than zinc. The affinity of the sensor for zinc should be high enough to bind at low concentrations, but not so high that it does not release zinc upon experiencing a significant decrease in zinc concentration. This optimal binding strength will allow measurement of changes in zinc concentration over time (flux); to be able to monitor flux, the sensor must also bind zinc in a rapid and reversible manner. For use within biological tissues, the ability to induce fluorescence by shining low-energy (visible to infrared) light is desirable, and it needs to be water soluble and stable over time. Ideally, it would also be nontoxic, which would allow it to be used in living organisms and potentially for medical applications.

Clearly, meeting all of these requirements presents quite a challenge! Nevertheless, there have been some impressive advances in this area. A number of zinc sensors have been developed, including

Trigger

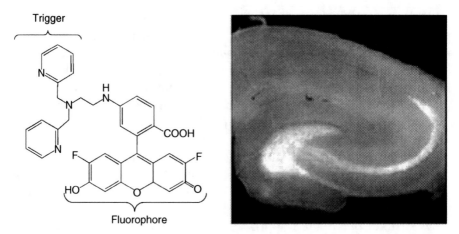

Fluorophore

**Figure 9.5.** *Structure of Zn(II) sensor ZnAF-2F showing fluorophore and trigger moieties (left). Fluorescence images of slices of the hippocampus from the brain of a rat loaded with a cell-permeable derivative of ZnAF-2F (right). The bright region indicates where zinc is concentrated. Adapted from Hirano, T.; Kikuchi, K.; Urano, Y.; Nagano, T. Improvement and Biological Applications of Fluorescent Probes for Zinc, ZnAFs.* Journal of the American Chemical Society *2002, 124, 6555–6562.*

some that display "turn-on" activity, meaning that they become fluorescent upon Zn(II) binding. The typical sensor design has a moiety consisting of a known fluorescent molecule and a trigger that can alter the sensor's fluorescence intensity when a metal binds (Figure 9.5). Some sensors have been designed that are taken up into cells, allowing intracellular zinc sensing. These types of sensors have been demonstrated to be effective at imaging zinc-rich regions in the brains of mammals (Figure 9.5). This approach to investigating inorganic species in the cell allows the inorganic chemist to delve into the intracellular environment, truly melding inorganic chemistry with biology.

## I'VE GOT METALS ON MY MIND

Now that I've given you a broad overview of the types of molecules studied and the techniques used in the broad field of bioinorganic chemistry, I want to tell you a little more about some specific areas of study within this field. I realize your interests are diverse, but you mentioned that applying chemistry to study biomedical problems appealed to you. Thus, I will focus on that area for my examples, which are the role of metal ions in AD and the use of metals as pharmaceuticals.

The recognition that metals play key roles in the central nervous system has spawned the field of "metalloneurochemistry," which in fact I already touched upon when I wrote about zinc sensing in brain tissue. Like all aspects of bioinorganic chemistry, this area itself is quite broad and includes studies such as the development of probes of neurologically relevant metals, the characterization of ion channel structure, the analysis of proteins involved in signal transduction by calcium in the nervous system, and the investigation of the effects of lead on the brain on the molecular level.

One neurological disorder in which metal ions play a prominent role is AD. The most common form of dementia (AD) is a progressive and fatal disease characterized by the formation of protein-containing plaques in the brain. These plaques consist of extended structures of repeating units of a peptide known as amyloid β (Aβ) and are called amyloid fibrils. The Aβ peptides form regular local structures known as β-sheets, which are a common feature of many soluble proteins. In the case of Aβ, however, these repeating structures form insoluble aggregates (Figure 9.6). There are many fascinating questions surrounding Aβ and AD. For one, although these plaques are always present in patients with AD, the role of the plaques in the disease progression is largely unknown. Second, the Aβ peptide is present in the brains of people who do not have AD, and its function is unknown, although there is evidence that it may play a role in memory or may have neuroprotective activity. Third, it has been shown that the Aβ peptide can undergo a conformational change known as misfolding, which can propagate the formation of amyloid fibrils, but what triggers this conformational change is not understood. Indeed, AD presents many interesting challenges for chemists to tackle.

So, where does inorganic chemistry come into the picture? It starts from observations that the levels of zinc and copper ions were abnormally high in the brains of AD patients. Subsequently, it was found that these metal ions can bind to Aβ in different ways, and zinc and copper may play a role in the propagation of amyloid fibrils by cross-linking peptides. More recently, it was reported that the nature of the groups from the Aβ peptide that bind to Cu(II) ions can play a role in propagating or inhibiting aggregation. When Cu(II) binds in one configuration, it inhibits aggregation, whereas binding in a different configuration accelerates aggregation (Figure 9.6). Accompanying studies on mouse neural cells revealed that the form of copper–peptide

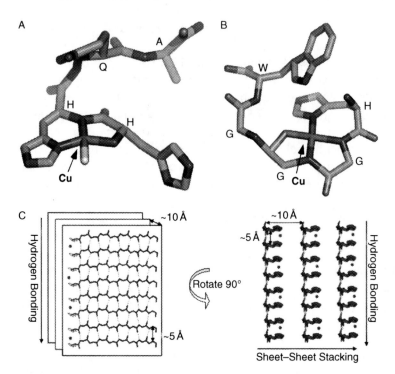

**Figure 9.6.** *Structural models for complexes between Cu(II) and peptides derived from the Aβ peptide sequence. (a) Proposed model for Cu(II) coordination with a peptide model for which Cu(II) binding inhibits fibril formation. (b) Crystal structure for a Cu(II)–peptide complex for which Cu(II) binding promotes fibril formation. (c) Structural model for peptide arrangements in fibrils. With the exception of one histidine side chain, only the peptide backbone is shown. The Cu(II) ions are depicted as small spheres. The dotted lines indicate hydrogen bonding between peptides arranged in a parallel β-sheet. Adapted from Dong, J.; Canfield, J. M.; Mehta, A. K.; Shokes, J. E.; Tian, B.; Childers, W. S.; Simmons, J. A.; Mao, Z.; Scott, R. A.; Warncke, K.; Lynn, D. G. Engineering Metal Ion Coordination to Regulate Amyloid Fibril Assembly and Toxicity.* Proceedings of the National Academy of Sciences of the United States of America *2007, 104, 13313–13318.*

that did not aggregate was nontoxic, whereas the form that aggregated was toxic. These results provide a direct link between the inorganic chemistry of metal–ligand interactions and a biological effect in the cell relevant to understanding AD etiology. Other studies in the field of bioinorganic chemistry that provide important background for this work include fundamental studies of copper binding to small ligands and peptides, and analyses of how metals impact protein and peptide structure and conformational changes. As our understanding of all of these phenomena continues to deepen, chances are increased for the development of a cure for this devastating disease. In addition, knowledge of how AD progresses may provide clues about how to prevent it from occurring in the first place.

# OPEN UP AND SAY "AU"

Studies of the roles of metal ions in AD fall within the field of "metals in medicine." In addition to analysis of how metal ions naturally present within an organism play a role in health and disease, this field also includes the study of the effects of metals introduced to the body, whether unintentionally (such as mercury in your tuna sushi) or intentionally in the form of metal-containing drugs. I thought I'd write a bit about metallopharmaceuticals, an area with a long history and continued promise.

The most successful metal-based drugs are cisplatin and carboplatin, platinum(II) complexes that are widely prescribed to treat a number of types of cancer, but particularly testicular and ovarian cancers. In my opinion, one of the more remarkable things about these drugs is how simple their structures are, with cisplatin being a quintessential simple coordination complex (Figure 9.7A). I remember that I was about your age when I initially heard of inorganic drugs, and the first thing I wondered was "however does someone first think to take a platinum compound to treat cancer?" Indeed, the story of how the anticancer activity of cisplatin was discovered is quite interesting and also provides a beautiful illustration of how science often gets done in "the real world"; for this reason, I believe it is worth summarizing for you here.

The discoverer of cisplatin's anticancer activity was Barnett Rosenberg, a professor of chemistry and biophysics at Michigan State University. What is interesting, although not unusual, about his story is that his research was not directed in any way toward anticancer drug development or even platinum complexes. Rather, his interest was in learning the effects of applying an electric field on bacterial growth. When performing this experiment, he noted that the

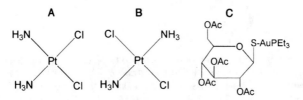

**Figure 9.7.** Structures of (A) cis-Pt(NH$_3$)$_2$Cl$_2$ (cisplatin), (B) trans-Pt(NH$_3$)$_2$Cl$_2$, and (C) auranofin.

*Escherichia coli* bacteria he grew in the presence of an electric field became highly elongated relative to normal *E. coli* because of the inhibition of cell division. Because he did careful control experiments, Professor Rosenberg determined that it was not the electric field but rather a compound produced during the experiment that had the effect on the bacteria. It turned out that the platinum electrodes were reacting with components of the bacterial growth medium to form platinum(II) compounds. Further experiments showed that it was the platinum complex with the formula $Pt(NH_3)_2Cl_2$ responsible for the activity, and specifically the isomer *cis*-$Pt(NH_3)_2Cl_2$ (Figure 9.7A), also known as cisplatin. Notably, the other isomer *trans*-$Pt(NH_3)_2Cl_2$ (Figure 9.7B) did not have the same effect.

This discovery was very interesting in itself, but Professor Rosenberg did not stop there. He realized that the ability of cisplatin to arrest cell division could have therapeutic value against cancer and proceeded to test its activity against tumors in mice. These experiments were successful, and in 1971, cisplatin entered clinical trials in humans and to this day remains a important anticancer drug. To reach this end result, Professor Rosenberg had to not allow himself to follow simple assumptions and had to be willing to explore a new area of research. Studies on the activity of cisplatin and related compounds continue and have provided valuable information on how it works on the molecular level, including how it binds DNA to cause it to bend from its helical axis (Figure 9.8). This fundamental knowledge provides a valuable basis for the development of other anticancer drugs.

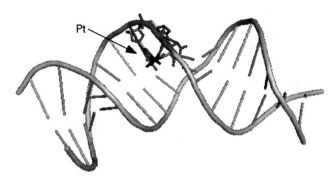

**Figure 9.8.** Structure of the adduct formed between cisplatin and DNA, determined by X-ray crystallography. The platinum is shown in black, and the two DNA bases that bind the platinum are shown in sticks. Note the bending of the DNA from its normally straight helical axis. Structure: Ohndorf, U. M.; Rould, M. A.; He, Q.; Pabo, C. O.; Lippard, S. J. Basis for Recognition of Cisplatin-Modified DNA by High-Mobility–Group Proteins. Nature *1999*, 399, 708–712.

Another class of inorganic drugs that have found wide clinical use are gold (in the form of Au(I)) compounds employed to treat the autoimmune disease rheumatoid arthritis. Although gold drugs such as auranofin (Figure 9.7C) have shown some success in reducing symptoms, their use has fallen out of favor in recent years because treatments that take effect more quickly and display fewer side effects have come on the market. Gold complexes are still used, but only as a last resort. Hampering the improvement of gold drugs has been that the means by which they act against rheumatoid arthritis has remained a mystery, at least until recently. Early in 2008, it was reported that gold salts block the release of a protein known as HMGB1 from the cell nucleus, a process that plays a role in promoting inflammation, a hallmark of the disease. It does this by inhibiting two other molecules, the protein interferon beta and the small molecule nitric oxide (NO), which are agents that promote HMGB1 release. Interestingly, there is significantly more HMGB1 in the fluid that protects joints than in other tissues in the body. This finding has reawakened interest in gold salts as treatments for this disease, as now that the action of gold is better understood, it may be possible to engineer faster-acting, safer, gold-based treatments. Improvements in drugs may come in the form of gold administered and/or the method of delivery. Additional advances that would help would be to obtain more molecular-level information on how gold interferes with molecules involved in HMGB1 release. This work is a good example of how studying how molecules work on a fundamental level has the potential to have a positive impact on people's lives.

The platinum and gold compounds I've written about here are only a few examples of inorganic drugs. Others that are in current use include gadolinium and technetium complexes that serve as contrast agents for medical imaging, iron-containing bleomycin, which is an anticancer agent, lithium salts for the treatment of bipolar disorder, and calcium and magnesium salts, which are used as antacids. Recent developments include the merging of metals in medicine and nanochemistry (or supramolecular chemistry), which falls under the umbrella of "nanomedicine." Of particular interest is the use of inorganic nanoparticles as drug delivery agents and as imaging agents. You should ask Professor Cohen there at UCSD more about these since he is an expert in this area.

The use of metals in medicine is an attractive area to those who like the idea of developing drugs that will alleviate symptoms and cure disease, that is, of having a direct and positive impact on people's lives. It is important to keep in mind that the road to developing a new drug is long and, as illustrated by the cisplatin story, is typically paved with serendipity. But any studies that enhance our understanding of how metals interact with biomolecules, how cells and tissues handle both native and introduced metal ions, and how naturally present metals contribute to healthy cell function will enhance our fundamental knowledge base that is essential not only for drug development but also for understanding how diseases develop and progress. In the end, any area of research that fascinates you is an area in which you will excel and will be a means by which you can make an impact on our world. It is indeed exciting and inspiring to undertake a career in which you are making a contribution to science in the broadest sense and thus to humanity. This holds for whatever area you decide to pursue. I hope you will consider bioinorganic chemistry as a specialty field within chemistry, as in my admittedly biased opinion, it offers great opportunities and an exciting range of problems to study and methods to apply. I wish you the best in your future career, and I expect I will see you again, probably at a conference, possibly in an exciting, faraway place.

Best wishes,

Kara

## FURTHER READING

Bertini, I.; Gray, H. B.; Stiefel, E. I.; Valentine, J. S. *Biological Inorganic Chemistry: Structure and Reactivity*, University Science Books, Sausalito, CA, 2007.

Bioinorganic Chemistry: Biocatalysis and Biotransformation. *Current Opinion in Chemical Biology* 2007, 7(2), 113–240.

Bioinorganic Chemistry Special Feature. *Proceedings of the National Academy of Sciences of the United States of America* 2003, 100(7), 3562–622.

Metals: Impacts on Health and the Environment. *Science* 2003, 300, 925–47.

Metals in Chemical Biology. *Nature Chemical Biology* 2008, 4(3), 143–57; 168–75; 185–93.

# 10

# *Better Than Sliced Bread*

**Chaitan Khosla**
*Stanford University*

Chaitan Khosla is the Wells H. Rauser and Harold M. Petiprin professor at Stanford University in the Departments of Chemistry, Chemical Engineering, and, by courtesy, Biochemistry. He received his PhD in 1990 at Caltech. After completing postdoctoral studies at the John Innes Centre in the United Kingdom, he joined Stanford in 1992. Over the past two decades, he has studied modular enzymes that make a medicinally important class of natural products called polyketides and has exploited their properties for engineering novel antibiotics. More recently, he has investigated celiac sprue pathogenesis with the goal of developing therapies for this widespread but overlooked disease. He has coauthored over 250 publications and 50 U.S. patents, and is the recipient of several awards and honors including the Eli Lilly Award in Biological Chemistry, the Pure Chemistry Award, and the Arthur C. Cope Scholar Award from the American Chemical Society, and the Alan T. Waterman Award from the National Science Foundation. He was elected Fellow of the American Association for Advancement of Science in 2006, a member of the American Academy for Arts and Science in 2007, and a member of the National Academy of Engineering in 2009. In 1995, he cofounded Kosan Biosciences, a public biotechnology company

*Letters to a Young Chemist*, First Edition. Edited by Abhik Ghosh.
© 2011 John Wiley & Sons, Inc. Published 2011 by John Wiley & Sons, Inc.

that developed new polyketide antibiotics. He is also a founder and director of Alvine Pharmaceuticals, a company that is developing an oral enzyme drug discovered in his laboratory for the treatment of celiac disease.

Dear Angela,

One of the most exciting frontiers for chemists lies at the interface between medicine and chemistry. Here I will focus on a subset of problems at this interface—the so-called orphan diseases. My goal is to persuade you that this grab bag of diseases presents an exceptionally fertile opportunity for new generations of chemists.

Let me start by addressing the three questions that immediately come to mind: (1) what is an orphan disease? (2) why should you care? and (3) how can chemistry make a difference?

## WHAT IS AN ORPHAN DISEASE?

Although the term orphan disease means many things to many people, it has a relatively well-defined meaning in the biomedical community. For example, in the United States, it is defined as a disease that affects fewer than 200,000 citizens. This includes extremely rare diseases, such as mad cow disease, which is (thankfully!) so rare that only a handful of cases have been diagnosed in the past several decades. It also includes many widespread tropical diseases, such as malaria, which are not particularly prevalent in the United States.

There are at least two reasons why these problems deserve your attention. First, almost by definition, orphan diseases are untrodden territory for research. The pharmaceutical industry is unlikely to work on these conditions because they provide little financial incentive. Biologists are also unlikely to pursue them because, with the exception of a few Mendelian diseases (single-gene disorders), animal models are lacking. Their low prevalence also makes them difficult to study in humans. So, between the test tube and the patient, there is an enormous gap, and it would not be an exaggeration to say that the health (and sometimes even the survival) of a patient with an orphan disease depends on the quality of molecular insight that emerges from the test tube. As you know well by now, test tubes are where chemistry happens. Take, for example, Gaucher's disease, a serious inheritable disease that affects many organs of the body. Until recently, only supportive therapy involving analgesics and surgery could be offered to patients. Today, an enzyme called imiglucerase, produced by recombinant DNA techniques (methods for making large quantities of a desired protein), is capable of reducing most symptoms of patients with Gaucher's disease; it's as good an example as any of the power of test tube chemistry. Second, this is an immensely rich and diverse problem area. Even though the prevalence of any given disease is low by definition, there are over 5000 different orphan diseases, most of which lack clinically effective tools for diagnosis, management, treatment, or cure. This means that there are millions of patients, with diseases like Gaucher's disease, waiting for a cure that will change their lives. In this age of molecular medicine, you could spend an entire lifetime cherry-picking your way through problems that motivate you without ever feeling like you're working in an intellectually overcrowded area.

So you're ready to sign on to a lifetime of investigations into orphan diseases. As a chemist, where do you start? That depends. If your attention has already been drawn to a specific orphan disease through a personal connection, you could start by learning more about it from the biomedical literature. On the other hand, if you don't have a specific pet project in mind yet, then check out the Web sites of the National Organization for Rare Diseases (http://www.rarediseases.org) or the National Institutes of Health (NIH) Office of Rare Diseases (http://www.rarediseases.info.nih.gov). For starters, I recommend getting a feel for the clinical face of a disease that interests you. Go talk to a physician. Get to know some patients and their family members. Because of their low prevalence, most patients with orphan diseases are seen by specialists in academic institutions rather than by community practitioners. This has both pros and cons. A potential disadvantage is that there may not be a local expert on your disease of interest (although at the University of California, San Diego [UCSD], chances are that there is one for virtually every disease!). The advantage is that a lot of knowledge about your disease of interest may be concentrated at these centers, so it may be relatively easy to access. The better you understand the actual nature of the problem (not merely your perception of the problem), the more empowered you will be as you dive into the molecular fundamentals of the problem to look for patterns, causal relationships, and solutions.

At a molecular level, a cell, tissue, or organ is astoundingly complex, and it is said that one does not truly understand a biological phenomenon until one can reconstitute its essence in a test tube with defined molecules. This is ground zero on the chemistry–biology interface. Chemists who are interested in biology seek to mimic the fundamental behavior of organisms and cells in the properties of individual molecules and their interactions with each other. This ability also allows them to reconstitute the essence of a disease in a test tube by studying how biomolecular behavior goes awry. In the case of an orphan disease, starting from even the most rudimentary hypothesis, chemistry has the power to create a molecular trail. The molecular logic of disease biology typically involves an atypically abundant biomolecule, a deficient biomolecule, a mislocated biomolecule, or a structurally defective biomolecule. Pinpointing and characterizing the cause and nature of any such error in a test tube is quintessential chemistry and is often the first step toward developing new diagnostics or therapies for a human disease.

Let me illustrate some of the above comments using my own experience with one orphan disease—celiac sprue. Celiac sprue (a.k.a. celiac disease, coeliac disease, gluten-sensitive enteropathy, or nontropical sprue) is an inflammatory disease characterized by poor absorption in the intestine, abnormal small-intestinal structure, and intolerance to gluten, which is a complex mixture of nutritionally important proteins found in common dietary food grains, such as wheat, rye, and barley. Although the disease was considered uncommon until recently, several epidemiological studies suggest that the prevalence of celiac sprue is in the range of 0.5–1.0% in most parts of the world. Like other immune disorders such as type 1 diabetes, rheumatoid arthritis, and multiple sclerosis, both genetic and environmental factors play a role in the onset of celiac sprue. Unlike these diseases, however, the expression of celiac sprue is dependent on dietary exposure to gluten. Patients enter into remission when they are placed on gluten exclusion, and they relapse when gluten is reintroduced into their diet. Celiac sprue is therefore unique among the chronic inflammatory diseases in that a critical environmental factor has been identified.

The disease commonly presents in early childhood with severe symptoms including chronic diarrhea, abdominal distension, and "failure to thrive." The general condition of these children is severely impaired. In many patients, however, symptoms may not develop until later in life when the disease brings about fatigue, diarrhea, weight loss, anemia, and neurological symptoms. This is a lifelong disease and, if untreated, increases the risk of complications such as bone disorders, infertility, and cancer. There is no therapeutic option available to celiac sprue patients; the only treatment is life-long adherence to a strict gluten-free diet. Complete gluten exclusion is very difficult to maintain. Besides its widespread use as a nutritional substance as well as an additive to enhance food characteristics or processability, gluten is an unlabeled ingredient in most packaged, bottled, and canned foods. Certified gluten-free products are not widely available and tend to be considerably more expensive than their non-gluten-free counterparts. Unsurprisingly, several recent clinical studies also suggest that intestinal malfunction persists in many patients in spite of efforts to exclude gluten from their diet. There is an urgent need to develop safe and effective therapeutic alternatives to a strict lifelong gluten-free diet for celiac sprue patients.

## ON THE HOME FRONT

I developed an interest in celiac sprue in 1999, when my 3-year-old son was eventually diagnosed with the disease after prolonged illness and poor growth. I quickly learned three lessons about the disease, only two of which were from the scientific literature. First, I started to realize how widespread and yet hidden gluten is in everyday life. It's not only there in breads and pastas but also in all kinds of unexpected processed foods and consumer goods (soy sauce and mailing envelopes are just two examples). Second, the literature gave me an appreciation for the hereditary nature of the disease. That, in turn, prompted my wife and me to get tested. To our dismay, we learned that my wife also had full-blown celiac disease. Today, even a small amount of contaminating gluten in their meals (unfortunately not an uncommon experience, especially when one eats outside of home) will make my wife and son violently ill. To think that she went through nearly three decades of life on a normal diet with such a serious immunological condition gives one pause. Third, and perhaps most pertinent to this letter, it struck me as odd that, although celiac sprue is perhaps the only human autoimmune disease (i.e., a disease in which one's immune system damages itself instead of attacking foreign objects) that can be reversibly triggered by a well-defined albeit complex chemical (gluten), it had been completely overlooked as a research opportunity by chemists.

A couple of days after we received the news about my wife's diagnosis, I got a call from Dr. Rita Colwell (then the Director of the National Science Foundation) informing me that I had been selected as the recipient of the 1999 Alan T. Waterman Award. One of the perks of this prize is a generous unrestricted grant for 3 years. Almost instantly, I made up my mind to use the funds to investigate celiac sprue. Not only would it enable me gain a deeper understanding of a problem that was here to stay in my family, but it would also give me a chance to educate some of the best and brightest students about the relevance of chemistry to the study of immune diseases.

## THE ROLE OF CHEMISTRY

Our approach to studying celiac sprue was motivated in part by the recognition that, among human autoimmune diseases, it is uniquely well suited to developing solutions through chemistry. As summarized above, the causative chemical, gluten, has been unambiguously identified. We therefore speculated (and still believe) that it should be possible to identify the Achilles' heels of celiac sprue by simply following the gluten trail. Broadly speaking, in our laboratory, we pursue the discovery and development of two types of therapeutic modalities. In one approach, we seek to detoxify gluten before it damages the primary affected organ (the small intestine). In a complementary strategy, we seek to block one or more nonessential human proteins that mediate the earliest stages of the pathogenic response to dietary gluten in a patient. In the remainder of this letter, I will give you a brief status report for each approach to illustrate how chemistry plays a central role in our efforts to translate scientific insight into practical treatment for patients.

First, a brief and simple background into protein chemistry. Proteins are polymers composed of 20 nutritionally important amino acids, strung together like beads on a string. Each amino acid is denoted by a letter of the alphabet. The chemical structures of three of these amino acids are shown in Figure 10.1 because they are especially important in the story that follows. For example, a bag of flour that you purchase from the grocery store has hundreds of gluten proteins, all of which share the property that they are rich in P and Q building blocks. The amino acid sequences of two such gluten proteins are shown in Figure 10.2.

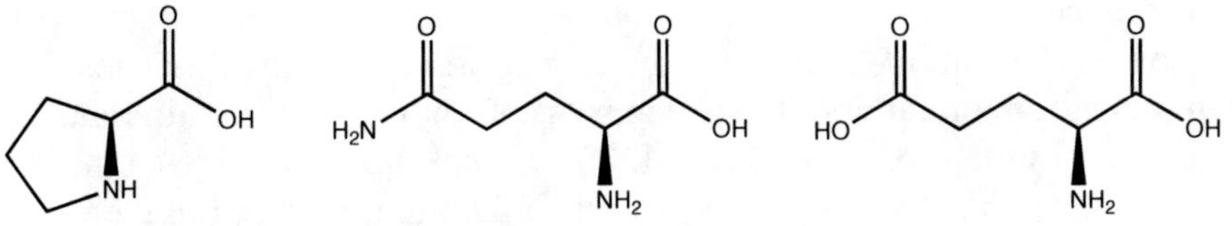

Proline (P)              Glutamine (Q)                    Glutamic acid (E)

**Figure 10.1.** *Chemical structures of proline (P), glutamine (Q), and glutamic acid (E).*

MNIQVDPSSQVQWPQQQPVPQPHQPFSQQPQQTFPQPQQTFPHQPQQQFP
QPQQPQQQ**FLQPQQPFPQQPQQPYPQQPQQPFPQ**TQQPQQLFPQSQQPQQ
QFSQPQQQFPQPQQPQQSFPQQQPPFIQPSLQQQVNPCKNFLLQQCKPVSLV
SSLWSMIWPQSDCQVMRQQSCQQLAQIPQQLQCAAIHTVIHSIIMQQEQQQG
MHILLPLYQQQQVGQGTLVQGQGIIQPQQPAQLEAIRSLVLQTLPTMCNVYVPP
ECSIIKAPFSSVVAGIGGQYR

**Figure 10.2.** *Amino acid sequences of two gluten proteins, α2-gliadin (top) and γ5-gliadin (bottom). Note the abundance of P and Q residues. The bold-faced sequences represent peptides that are resistant to digestive breakdown. See text for details.*

A series of experiments performed during 2000–2005 in collaboration between my laboratory and that of Ludvig Sollid, a pioneering Norwegian immunologist, revealed the structural basis for gluten toxicity in celiac sprue. Specifically, our research showed that representative gluten proteins such as α2- and γ5-gliadin (Figure 10.2) are difficult to digest by the mammalian gastrointestinal tract. When you eat protein in your diet, individual protein molecules are broken down, first into shorter chains of amino acids called "peptides" and ultimately into the amino acids themselves, by a process known as "proteolysis." The resulting amino acids are the actual substances that are absorbed by the human body. Proteolysis of a dietary protein begins in the stomach but mostly occurs in the upper small intestine through the action of enzymes called "proteases." The overall process is incredibly efficient in the case of most common dietary proteins, but gluten proteins are a notable exception. Gastrointestinal proteolysis of gluten proteins yields peptides that are unusual in four respects. First, they are considerably longer than the average peptide generated from other dietary proteins. Second, their subsequent breakdown into amino acids is extremely inefficient. As a result, these gluten peptides persist in intact form through the upper small intestine. Third, they are recognized by another enzyme present in intestinal tissue called transglutaminase 2 (TG2), which converts selected Qs in these peptides into Es. (Look at the chemical difference between Q and E in Figure 10.1; perhaps that will give you a sense of what TG2 does to gluten peptides.) Finally, this chemical modification enhances the propensity of these peptides to bind to another human protein, HLA-DQ2. Once these peptides bind to DQ2, the game is over. A vigorous inflammatory response ensues, leading to intestinal damage, malnutrition, and inflammation throughout the body.

A vivid example of such a peptide is the 33-residue peptide derived from α2-gliadin,LQLQPFPQPQLPYPQPQLPYPQPQLPYPQPQPF (Figure 10.2). The mechanisms underlying its formation and inflammatory character are depicted in Figure 10.3. An entirely analogous response is elicited by the 28-residue peptide from γ5-gliadin shown in Figure 10.2.

As the fundamental findings summarized above started to emerge, so too did the prospect of a simple but potentially powerful therapeutic strategy for treating celiac sprue. We hypothesized that, if oral enzymes could be identified to rapidly digest immunotoxic gluten peptides like the 33-mer while food is churning in the stomach, it would be possible to detoxify dietary gluten before the meal reaches the small intestine. For reasons outlined above, we were interested in enzymes that cleaved at Q and P residues in Q- and P-rich gluten proteins. Such enzymes have now been discovered from a variety of microorganisms and plants. Two of them, including a Q-specific protease from the barley plant and a P-specific protease from a bacterium, have recently entered human clinical trials. Although it will be several years before their clinical utility can be validated, I hope this gives you a flavor of how simple chemical insights can be translated into clinical strategies.

If you think about the scheme shown in Figure 10.3, you may recognize that a complementary approach to protecting a patient from the toxic effects of dietary gluten is to block the action of TG2, which in turn might shut down gluten-induced inflammation. Back in 1999,

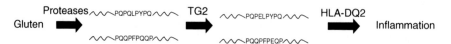

**Figure 10.3.** *Key chemical events in the pathway by which dietary gluten proteins induce inflammation in the small intestine of a celiac disease patient. Peptides such as the 33-residue sequence from α2-gliadin and the 28-residue from γ5-gliadin are shown as squiggly lines with a portion of their sequences explicitly listed. Transglutaminase 2 (TG2) converts selected Qs into Es in these peptides. The resulting peptides bind to HLA-DQ2 and induce inflammation. For details, see text.*

we postulated that TG2 might be a safe and effective drug target for the treatment of celiac disease on the basis of two observations. First, most gluten peptides that elicit an inflammatory response in a patient's gut are recognized and modified by TG2; this modification is a crucial early event in the adverse reaction of the patient to dietary gluten. Therefore, if TG2 could be blocked while the food is passing through the gut, one may have an effective therapy. Second, genetically engineered mice, in which the TG2 gene is knocked out, had just been reported in the literature. They were physiologically and reproductively normal, suggesting that local inhibition of TG2 in gut tissue may not present long-term health risks to patients. Together, these two observations suggested that blocking TG2 could be a safe and effective way to protect a celiac patient from the harmful effects of gluten. While this hypothesis remains to be clinically verified, chemistry will once again provide the critical tools.

Over the past 5 years, we have been working on developing a small-molecule inhibitor of TG2 that can be used as an experimental drug in humans. Simultaneously, we are trying to develop an animal model for such a clinical candidate. We have explored three different classes of synthetic TG2 inhibitors and have developed both a rodent and a monkey model to test these compounds. In the next 2–3 years, we hope to bring one such inhibitor into the clinic for proof-of-concept testing. There are three possible outcomes of such a study. On one hand, we may learn that, even when intestinal TG2 is blocked, gluten-induced disease cannot be prevented. On the other hand, we may learn that our small-molecule inhibitor can provide clinical benefit to the patient under certain controlled scenarios. In that situation, we will have both a drug candidate and a clinically validated target, and it will be a matter of time before such a drug reaches the marketplace. In yet another scenario, it is possible that our experiment will lead to the validation of TG2 as a drug target, but our drug candidate turns out to be associated with other unrelated problems, in which case it's back to the chemical drawing board to find a better mousetrap.

I hope I have given you a feel for the relevance of chemistry in orphan diseases. In the case of celiac sprue, it has been half a century since W. K. Dicke, a Dutch pediatrician, identified gluten as the principal environmental trigger of the disease. During this period, the prevalence and etiology of this disease have been extensively investigated by biologists and clinicians, but it has been completely overlooked as a therapeutic opportunity. I firmly believe that chemistry can fill this gap in the next decade. Celiac disease is by no means unique across the spectrum of orphan diseases. Take for instance enteric infections that cause diarrhea. While thankfully rare in the United States, such infections cause 1.6–2.1 million deaths a year throughout the world, mostly in children under the age of 5 in the developing world. Surely your generation has the willpower and brains to solve this problem. If so, chemistry would certainly be an excellent place to start. I encourage you to consider this avenue of advanced study in your own career.

Best wishes,

Chaitan Khosla

## FURTHER READING

Bethune, M. T.; Khosla, C. Parallels between pathogens and gluten peptides in celiac sprue. *PLoS Pathogens* 2008, *4*, e34.

Fasano, A. Surprises from celiac disease. *Scientific American* 2009, *301*, 54–61.

Maeder, T. *The Orphan Drug Backlash*, Scientific American, 2003, pp. 81–87.

Petri, W. A.; Miller, M.; Binder, H. J.; Levine, M. M.; Dillingham, R.; Guerrant, R. L. Enteric infections, diarrhea, and their impact on function and development. *Journal of Clinical Investigation* 2008, *118*, 1277–1290.

# 11

## Choreographing DNA

**Cynthia J. Burrows**
*University of Utah*

Cynthia J. Burrows, Distinguished Professor of Chemistry at the University of Utah, holds a BA from the University of Colorado and a PhD from Cornell University and was a postdoctoral fellow with Nobel Laureate J. M. Lehn in Strasbourg, France. She was a faculty member at State University of New York at Stony Brook prior to moving to Utah in 1995. Her research interests are in bio-organic chemistry with a focus on DNA and RNA. She and her physical chemist husband, Scott Anderson, have three children (triplets) of whom Laurel is the youngest, by a minute.

Hi Angela,

The dress rehearsal went smoothly last night; thanks for asking. For their annual concert, the University of Utah's Children's Dance Theatre selected topics in science around which to create an original dance and music performance entitled *Scientia*. Each class choreographed a dance highlighting a different discipline ranging from biology and medicine to physics and electronics. The story line followed the journey of a young woman, Asha, accompanied by a

*Letters to a Young Chemist*, First Edition. Edited by Abhik Ghosh.
© 2011 John Wiley & Sons, Inc. Published 2011 by John Wiley & Sons, Inc.

*165*

wise but eccentric (aren't they all?) old professor traveling in his time machine to visit the past, present, and future of our planet and universe. The dances helped the viewers learn about viruses, lasers, rain forests, the earth's core, and Newton's *Principia Mathematica*, and the setting ranged from the microscopic to, literally, the galactic.

Our daughter Laurel is in tenth grade this year, so she was dancing with the High School I class directed by Amy Daly. Astronomy has been a passion of Laurel's from an early age, and so when her class was given the topic of "galaxies" to choreograph, she was really delighted. The class spent the better part of 6 months learning about galaxy shapes, formation, and motion, and then trying to convey these concepts into a dance format. Early on in the process, I had suggested to Mary Ann Lee, the artistic director of Tanner Dance, that she should recruit experts from the science departments at the University of Utah to add scientific authenticity to the dances being created. Laurel's class invited Professor Paulo Gondolo, from the Department of Physics, who is an expert in cosmology and dark energy. He showed stunning photos from the Hubble Space Telescope illustrating the differences between spiral, elliptical, and other classes of galaxies. At the dress rehearsal last night, the dancers' swirling violet skirts studded with stars added to the imagery of various galactic shapes and motions in the dance.

It was my job to be the volunteer "expert" on DNA for the ninth-grade class, directed by Tina Misaka. My lab has been studying the chemistry of DNA and RNA for nearly 20 years, but we are far from understanding all the intricacies of these vital biomolecules. Nevertheless, we had a great discussion about some of the current knowledge of DNA and RNA structure, and particularly the dynamic aspects of DNA that are so important to its function. I was impressed that some of the ninth graders were able to ask such challenging questions about the medical outcome of changes in our genetic code. I think the highlights of what we discussed may interest you as well.

## A BIT OF HISTORY, BACKGROUND, AND BASICS

In 1953, James Watson and Francis Crick proposed the double helical structure of DNA based on X-ray crystallographic data obtained by Rosalind Franklin (Sayre, 1975). The double helical form has long been admired in architecture as a graceful and intriguing form (see

A                    B

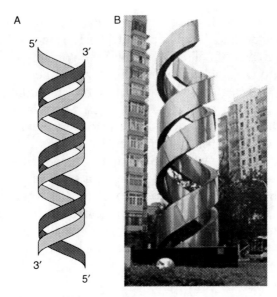

**Figure 11.1.** (A) A 2-D drawing of the DNA double helix on paper. (B) A 3-D sculpture of a double helix in downtown Beijing.

Figure 11.1), but when constructed in wood, glass, or steel, it is rigid and immobile. That is not at all the case for DNA! As a molecule, it is constantly in motion. And it has to be, because it is the storehouse of information for the genetic code. Imagine, if all books in a library were glued to the shelves, we could never read their contents. Thus, it is particularly appropriate to represent DNA as a dance where its changes in form choreograph processes in the cell.

The DNA molecules in each cell are very long; this is one of the first concepts about DNA structure that illustrates why it cannot be a simple, twisted rigid rod. If the DNA in your 46 chromosomes were lined up end to end as extended double helices, its length would be about 2 m, or more than 6 ft. How could this possibly fit inside a cell nucleus whose diameter is about a tenth of a millimeter? DNA has to bend. And it bends quite cleverly. Consider this: the genomic sequence is made up of the four bases guanine (G), adenine (A), thymine (T), and cytosine (C) linked together by a sugar–phosphate backbone. The order of the letters, GATC ... or perhaps CGCGTAGTAAC ..., codes for the order of amino acids that will be placed together in a polypeptide to make a specific protein. Some of the sequences are relatively rigid and don't want to bend; others are adept at making slight deformations or even slippages that allow the helix to twist into a supercoil. This coiling allows the DNA to wrap around a set of core histone proteins that package the DNA helix into nucleosomes like beads on a string. Further coiling continues until you have coils of coils of coils, and the condensed chromosomes actually fit into the nucleus. All of this happens because slight variations in the DNA sequence permit a tiny bit of bending, and if these bending sites happen once every 10 bases, which is one turn of the helix, then they add up cooperatively to become a big bend and eventually circular loops, as in Figure 11.2.

Now, what happens when we want to "read" a sequence of the genome that is buried inside this coil of coils? If we were in a library, we would have to go the right section, find the right shelf, find the right book, open to the right chapter, and begin to copy down the part we are interested in. The dynamics of how all this happens within the cell is very much at the frontier of current research in DNA.

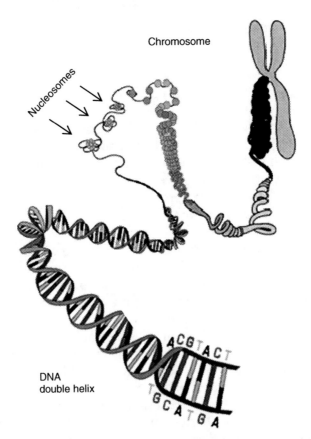

**Figure 11.2.** *The DNA duplex wraps around histone proteins to form nucleosomes that further coil and coil to become tightly packaged chromosomes.*

Specific proteins perform this function by restructuring the chromo-somes to provide access to the parts of interest.

Markers may also be placed on the DNA, sort of like bookmarks. For example, a single carbon in the form of a methyl group ($CH_3$) can be added to cytosine to slightly perturb the shape of the DNA at that site, so that proteins searching through the DNA will recog-nize that section as different next time. This process of DNA meth-ylation is part of the blossoming field of epigenetics, a sort of supercode frosted on top of the A,T, G, C base sequence that turns off certain genes that are no longer needed—say, later in life. Continuing the process, we still need to find our gene of interest, so the DNA helix rolls and loops along on the nucleosomes like an inchworm, in order to expose different segments of DNA. Once the starting point of a gene is found, we are ready to transcribe the gene sequence and make RNA.

Transcription is the process of copying down a negative (or a com-plement) from the positive template of the DNA. Because DNA is made up of two strands, there is both a genetic code and a backup version opposite. Watson and Crick described this in their structure of the double helix with the concept of base pairing: Guanine (G) forms three hydrogen bonds with cytosine (C), and adenine (A) forms two hydrogen bonds with thymine (T) (see Figure 11.3). It is this complementarity that leads to a mechanism of copying DNA (Edelson, 1998). All the sequence information is contained in one strand of DNA; the other is a backup copy, albeit as a negative.

During transcription, we copy only the template strand of DNA by synthesizing one complementary strand of RNA called messenger RNA. RNA is based on ribose as the sugar rather than deoxyribose and utilizes uracil (U, lacking a $CH_3$) in place of T. These subtle

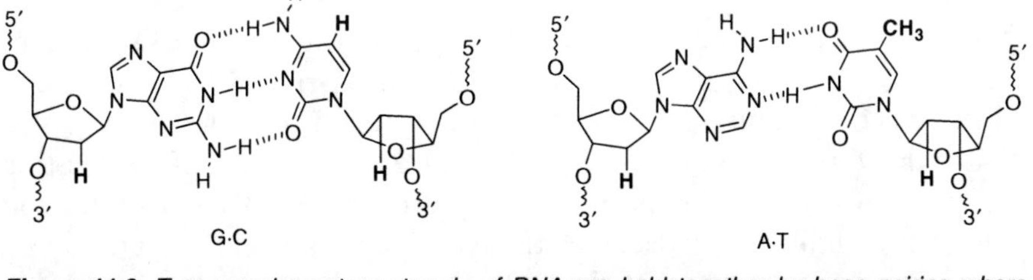

**Figure 11.3.** *Two complementary strands of DNA are held together by base pairing where G·C pairs make three hydrogen bonds and A·T pairs have two H bonds. For C, the H indicated in bold can be replaced by a $CH_3$ group yielding 5-MeC, an epigenetic marker. In RNA, the $CH_3$ group of T is replaced by an H (making U) and all the 2' positions (bold-faced H on sugars) carry OH groups instead of H.*

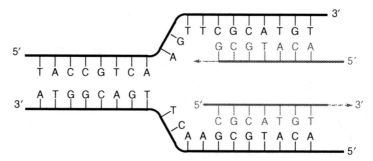

**Figure 11.4.** *Reading and copying DNA. The original strands are shown in black and the copies in gray. For transcription, only one strand (top) is copied as an RNA complement, G inserted opposite C, A inserted opposite T, and so on. For replication, both strands are copied as DNA complements. The top strand is synthesized continuously in the 5'–3' direction, while the lower strand is synthesized in segments and later stitched together.*

differences give the helix a different shape, which permits specialized functions in the cell. The mRNA heads off to the ribosome, outside of the nucleus, where its function is to provide the code to synthesize a sequence of amino acids in a new protein molecule.

The concept of copying DNA takes place also for replication in addition to transcription, the difference being that both strands of DNA serve as templates to make two complementary copies leading to a (hopefully) perfect copy of the genome (Figure 11.4). This is what you need to do just prior to cell division. The process of replication is fairly complex and requires the coordinated efforts of many proteins. First, you need to untwist and pull apart the two DNA strands. Then, because the strands are running in opposite directions (one is 5' → 3' and the other is 3' → 5', where the numbers refer to the two ends of the sugar connections), and because polymerase enzymes can only synthesize in the 5' → 3' direction, the two strands are copied in different ways. One strand is synthesized continuously (Figure 11.4, top), and the other strand requires that the DNA template be pulled out in loops to conduct the synthesis of a segment. These discontinuous segments are then stitched together later.

On the dance floor, the students had an easy time of illustrating complementary base pairs and coiling strands of DNA. Replication was more challenging because the two strands of DNA are doing different things at different times. With only about 20 students in the class, it is hard to demonstrate what a few hundred or thousands of DNA bases are doing in this elaborate dance called replication. Nevertheless, the dance managed to demonstrate that one could bring order to all these seemingly chaotic events.

# DNA DAMAGE, MUTATIONS, AND REPAIR

Research on DNA replication has come a long way since Watson and Crick. We now know that there are almost 20 different proteins in the human cell that have different roles to play as polymerases, or enzymes that put together the small monomers A, T, G, and C into a long DNA polymer in the correct sequence. To make an accurate copy of the $10^9$ bases in the human genome, DNA has to be copied without making more than 1 in $10^{10}$ mistakes! How is this possible? Well it's not, but polymerase enzymes do come close. Generally, the high-fidelity polymerases only make a mistake 1 in $10^4$ bases as they synthesize DNA. Most also contain a "proofreading" domain, which double checks the sequence just made to be sure the bases are complementary. If they don't match, the proofreader rips out the stitches and tries again. This gets your error rate down to about 1 in $10^7$. Then, DNA repair enzymes scan the duplex, particularly before transcription and before replication, to look for mistakes. If they find perturbations in the helix, out come the molecular scissors to snip out the bad part, and then a specialized polymerase puts in the right bases opposite the template. Altogether, the 3 billion base pairs in DNA can be copied *nearly* perfectly.

Is this good enough? Sure. We want some variation in our genomes after all; otherwise, we would all look like clones, and there would be no possibility for adaption and evolution. Unrelated humans differ from each other by about 0.1% of their DNA sequence; humans and monkeys are different by about 1%, and humans and rats by only a bit more. Finding the subtle ways in which human genomes differ, and particularly those differences that result in a genetic susceptibility to a disease, is a major goal of the human genome project directed toward personalized medicine. This area of research will expand beyond belief in your professional lifetime.

Errors made by polymerases account for only some of the changes or mutations observed in the genome. Sometimes, polymerases have no choice to but to guess which base to insert next because the template base has been chemically damaged in some way that makes it unrecognizable. Here are a couple of examples that come from *oxidative stress*, that is, the escape of oxygen free radicals of respiration and metabolism and their attack on DNA bases. Let's start with guanine, which is the most susceptible to oxidation. Its common oxidation product is 8-oxoguanosine (or OG) in which an oxygen atom has been added to carbon number 8 of the base (see Figure 11.5).

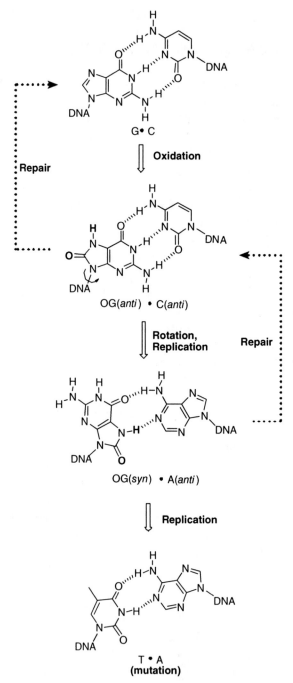

**Figure 11.5.** *A G · C base pair can undergo mutation to a T · A base pair via oxidation to 8-oxoguanosine (OG), 180° rotation, and misinsertion of A opposite OG.*

This doesn't look like it should change the base pairing to C, but during replication, the OG base can flip over to present a different set of atoms for hydrogen bonding. Now its complementary base is A instead of C. If the strand containing A undergoes replication, a complementary T will be placed opposite. This T would occupy the position that was originally a G. This is called a G → T transversion mutation. Oxidative events like this occur about 10,000 times per cell per day. If we couldn't correct these errors, we'd all mutate into slime mold by the end of the week! (Well, maybe not slime mold, but it wouldn't be pretty.)

DNA repair is a lifesaving process for the genome (David, 2007). Several parallel mechanisms operate to get rid of damaged bases that were injured by oxygen or nitrogen radicals, toxic substances in our food or air, or just an excess of a reactive species being in the vicinity of DNA at the wrong time. For example, base excision repair (BER) enzymes scan the genome for bases that don't fit, and then break the bond between the base and sugar so that other enzymes can come together to put in the correct base opposite the undamaged template. Accumulation of DNA damage that overwhelms the repair processes, or defective repair enzymes, can lead to serious diseases including cancer and aging. As DNA damage and mutations accumulate, the cell becomes less competent and ultimately might decide to die rather than replicate its faulty genome. In fact, manipulating DNA repair is something that chemists would really like to do. If you could find molecules that boost DNA repair, it would help prevent cancer and aging. Or, if you could deliver medicines to cancer cells that inhibit DNA repair, the cancer cell would die on its own. Here again is a wide open horizon for a young chemist interested in DNA and medicine.

## CHOREOGRAPHING DNA

The ninth-grade class and I spent quite a bit of time talking about DNA damage, mutations, and repair, and the group experimented with damaging and fixing their sequences. I'm not sure the casual observer could have deciphered all the complexities of this dance, but knowing the conversations underlying the choreography, I thought it was spectacular.

For costumes, Cynthia and Wendy Turner designed dresses of black mesh into which we wove strands of blue, yellow, green, and red ribbons, primary colors representing the four-base building blocks. Each pattern of color on the dress was unique, just like the individual wearing it! I enjoyed helping out in the costume shop; sewing the coarse mesh fabric was a bit challenging, but weaving the colorful ribbons was fun. I tried to pick out the patterns I'd constructed as the dancers whirled by on stage, but it all happened too fast.

Tristan Moore composed the music for each segment of the dance, and we experimented one day with sounds to go along with a DNA dance. While I was lecturing to the class about DNA structure and replication, Tristan was recording my voice. Later, Miss Tina asked the students to shout out key words from what we had discussed. I heard "coils of coils," "complementary base pairs," "your own personal blueprint," "6 billion bases," and "mutation." A few minutes later, Tristan played back what he had composed on a synthesizer, a melodic piece with my somewhat distorted voice echoing "coils of coils of coils of coils ...," and "mutations, mutations, mutations. ..." The dancers did a trial run with the music, and it was a magnificent first attempt. Of course, the final performance was finely tuned, the timing was perfected, the choreography was enriched to include both scientific concepts and artistic elements, and the costumes and lighting added to the success of the project.

My only regret is that I couldn't cover every topic in nucleic acid chemistry while I was talking to the dancers. My own area of research relates to DNA damage leading to mutations. We study the further oxidation of the OG base that I discussed above; OG is a hot spot for further oxidation, and it produces hydantoins, which are nearly 100% mutagenic in a bacterial cell assay. Thus, we really need to know how often these damages occur, when they occur, what the consequences are, and how to avoid them—take more vitamin E? Eat more carrots and kiwis? Personally, I vote for more chocolate in my diet!

Epigenetics, as I mentioned earlier, is an emerging field of DNA research in which the simple process of adding or subtracting one carbon, a methyl group, to a C base (to make 5-MeC) has a profound effect on whether or not that section of DNA is transcribed (Bird, 2007). However, the process is much more mysterious than originally thought. Methyl groups don't always silence a gene. For example, we think that one of the two copies of X chromosomes in women needs to be silenced, but recent studies have shown that an active X chromosome might have more methylation than its silent partner. This is very confusing and begs for more research.

## LET'S NOT FORGET RNA

Then there's RNA, which arguably raises many more scientific questions than DNA. RNA is often single stranded and loops back on itself to form hairpins, cruciforms, and pseudoknots. In recent years, a number of small RNAs have been found to function in the cell in powerful ways: small interfering RNAs (siRNAs) and micro-RNAs (miRNA) are capable of regulating protein synthesis, and this means that their introduction into the cell might be used in a medicinal sense.

Going back to the beginning of life, we are also learning more about how RNA was the primordial molecule that performed the modern-day functions of DNA and proteins (Yarus, 2010). Certain RNAs have catalytic or regulatory properties like proteins, and the RNA base sequence might even have been the original information storage system. Our lab is investigating how modern vitamins such as riboflavin evolved from RNA building blocks; we are attempting to create experiments that might model life processes on early Earth several billion years ago (Burrows, 2009). Lacking the time machine of the Professor in *Scientia*, we can only hypothesize how air, water, soil, and sunlight got together to form the first archaebacteria.

So, Angela, you can see that there is quite a lot of chemistry involved in the study of DNA and RNA, and so much more to be done. When I think of the importance of DNA in our lives and the amount of this material on Earth, I am awestruck. Each cell has one copy of the genome, those 2 m of DNA we talked about. Each of us has nearly $10^{14}$ cells in our body. There are $6 \times 10^9$ people on Earth. Altogether, this means there is 1 mole (Avogadro's number or $6 \times 10^{23}$) of the human genome on the planet. And, if you stretched that mole of DNA out, each molecule end to end, you would cover the amazing distance of nearly $10^{20}$ km—about 0.1% of the diameter of the universe! (Now, if only we could think of a way to use DNA unraveling/refolding for intergalactic travel?)

It's getting late and I need to stop. I do hope you have enjoyed this chemical tango with DNA, and that you will think of it when you study nucleic acids in organic and biochemistry courses at University of California, San Diego (UCSD). There will be yet more unimaginable frontiers ahead of you by then.

With my best wishes,

Aunt Cindy

P.S. I'm just back from opening night. When the curtain went up, I saw 18 dancers in undulating chains of DNA, not static, twisted ladders, but dynamic building blocks of the genome, constantly in motion: coiling and uncoiling, base pairing and reacting, and being repaired or chaperoned into a new task, all this set to music, composed, coordinated ... choreographed!

## ACKNOWLEDGMENTS

Special thanks to Miss Tina Misaka's ninth grade class for a marvelous experience. I would also like to thank Professor Mary Ann Lee (Artistic Director of CDT) and Ms. Laurel Anderson for advice and critical comments, as well as the National Science Foundation for financial support of our research program.

## FURTHER READING

Bird, A. Perceptions of epigenetics. *Nature* 2007, *447*, 496–498.

Burrows, C. J. Surviving an oxygen atmosphere: DNA damage and repair. In *Chemical Evolution II: From Origins of Life to Modern Society*, Zaikowski, L.; Friedrich, J. M.; Seidel, S. R. (eds.), ACS Symposium Series, American Chemical Society, Washington, DC, 2009.

David, S. S.; O'Shea, V. L.; Kundu, S. Base-excision repair of oxidative DNA damage. *Nature* 2007, *447*, 941–950.

Edelson, E. *Francis Crick and James Watson: And the Building Blocks of Life*, Oxford University Press, New York, 1998.

Sayre, A. *Rosalind Franklin & DNA*, Norton, New York, 1975.

Yarus, M. *Life from an RNA World*, Harvard University Press, Cambridge, MA, 2010.

*Part III*

# *Functional Materials*

# 12

# *Supramolecules to the Rescue!*

**Seth M. Cohen**

*University of California, San Diego*

Seth M. Cohen is an associate professor at the University of California, San Diego. He received his BS in chemistry and his BA in political science from Stanford University. He earned his PhD at the University of California, Berkeley, under the mentorship of Professor Kenneth N. Raymond, where he studied ligand systems relevant to bacterial iron transport and magnetic resonance imaging contrast agents. He was a National Institutes of Health (NIH) postdoctoral fellow in the laboratory of Professor Stephen J. Lippard (Massachusetts Institute of Technology [MIT]), where he performed studies on the interaction of transcription factors with cisplatin-damaged DNA. At the University of California, San Diego, the Cohen laboratory studies the design of metalloprotein inhibitors and the synthesis of functionalized metal–organic frameworks (MOFs).

*Letters to a Young Chemist*, First Edition. Edited by Abhik Ghosh.
© 2011 John Wiley & Sons, Inc. Published 2011 by John Wiley & Sons, Inc.

Dear Angela,

Welcome back to the University of California, San Diego—I hope the summer break didn't seem to pass too fast. I enjoyed having you in Chemistry 6CH Honors General and Forensic Chemistry last spring. If I recall, you had lunch with one of our guest speakers from the Drug Enforcement Administration, a rare treat that only a few of the students were lucky enough to participate in. I do remember that your final poster presentation for the class was terrific; I believe it was on an article that applied atomic absorption spectroscopy to detect heavy metals in hair samples.

How is your sophomore year going? I am sure that if you are showing the same tenacity and enthusiasm for your classes this year as you did in my honors freshman chemistry course, then you must be doing great! Your sophomore year is a good time to get a better sense of whether you want to major in chemistry or not. You will really delve into some of the "classic" coursework, namely, organic chemistry, which may spark your interest in the chemical sciences. Many students can decide after their first quarter of organic chemistry that they absolutely love, or hate, chemistry. I hope you will find this to be a similarly decisive experience—of course, I hope the encounter is favorable!

Another good thing to consider getting involved with this year is the American Chemical Society Student Affiliates (ACSSA) chapter. We have a really active ACSSA group that hosts a variety of chemistry-related student activities. These events can range from "meet the faculty" dinners to tours at laboratories of local pharmaceutical companies. The ACSSA also participates in the American Chemical Society (ACS) National Chemistry Week at Balboa Park by hosting a booth and putting on demonstrations for members of the public. The University of California, San Diego, ACSSA chapter always puts on a great show during these public events. One other nice thing about the ACSSA is it offers a Summer Research Fellowship, which you could be eligible for, to provide some financial support to work in a research laboratory this summer (maybe the Cohen laboratory, hint, hint). Anyhow, just some food for thought.

That said, let me tell you a little bit about one area of chemical research I work in, namely, supramolecular chemistry. I'll tell you about this topic in parts, first discussing some of the history of supramolecular chemistry, then describing a few of the current exciting areas of research, and finally writing a bit more about my specific interests in the field. Hopefully, this will give you a nice overview of the subject and lead you to read a bit more about this area of chemistry (I have provided a few references for additional reading at the end).

## CHEMISTRY BEYOND THE MOLECULE

Supramolecular chemistry has been described as "chemistry beyond the molecule." Although a very broad statement, it does generally summarize the basic concept, which is the study of how molecules can interact and associate with each other to generate more complex structures with more sophisticated functions. The original "supramolecular chemist" is Mother Nature, who has created the greatest achievements in supramolecular chemistry. A good example of supramolecular chemistry in nature is viruses. You probably know viruses more for their ability to cause diseases, like the common cold. But if you take a closer look at a virus, what you see is one of the most spectacular examples of supramolecular chemistry.

In its simplest incarnation, a virus is a "capsule." The outside of the capsule, called a "capsid," is made up of proteins, and inside the capsule is stored the genetic material of the virus (either DNA or RNA). The protein capsid is where we find our example of supramolecular chemistry. The capsid is composed of numerous (hundreds or thousands) individual proteins that self-assemble into a casing of specific size, shape, and symmetry. This is achieved by the combined effect of many different "weak interactions" that work in concert to generate the final structure. Different viruses have different capsid structures, demonstrating that nature has mastered the ability to make "molecular containers" of almost any shape or structure by using supramolecular chemistry. Perhaps, most amazing about the viral capsid is the specificity of the self-assembly process; that is, in the nearly limitless number of possible ways in which the protein "pieces" could come together (Figure 12.1), nature has designed each piece carefully such that only a single structure is obtained. Again, this specificity is achieved by the very precise design of numerous weak points that come together in just the right way.

By comparison, the efforts of humans to utilize supramolecular chemistry seem quite modest; however, we are getting better at it all the time. Also, the fact that supramolecular chemistry has not

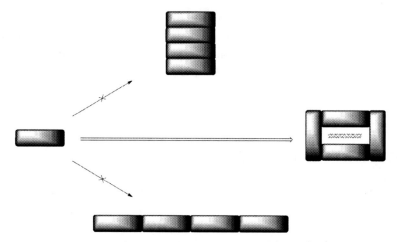

**Figure 12.1.** *Simplified scheme of the self-assembly of four capsid "proteins." The four proteins (shown as filled rectangles) could arrange themselves in many different ways, but they only form one structure, based on the combination of many different weak interactions, to form a capsule to hold the genetic material (squiggles) of the virus.*

completely matured as a field means that young scientists like yourself are poised to make the greatest discoveries in the years to come! Some of the earliest efforts in supramolecular chemistry started with a concept not that dissimilar from that of a virus—the idea was simply to design a molecule that could encapsulate a smaller molecule (just as the virus has a capsid to encapsulate the genetic material). This concept is often referred to as "host–guest" chemistry, where the host molecule binds to and encases the guest molecule. There are many simple analogies that are suitable to describe this kind of supramolecular arrangement, but perhaps now is the time to introduce the comparison you may have been anticipating—the "supramolecule." A seemingly "mild-mannered" molecule, the supramolecule can grab hold of the guest and "protect" it from its surroundings. We will return to this silly analogy later, as it is really quite relevant for one of the early types of supramolecular systems.

Some of the first host–guest supramolecular systems involved the encapsulation of metal ions (charged metal atoms) inside cyclic organic molecules ("macrocycles"). In these supramolecules, the metal ion sits inside the center of the macrocycle and is held in place by weak interactions between the metal ions and heteroatoms ("heteroatoms" refers to the noncarbon and nonhydrogen atoms in the macrocycle). In this case, these weak interactions can be described, in a very basic sense, as the attraction of opposite charges. Metal ions

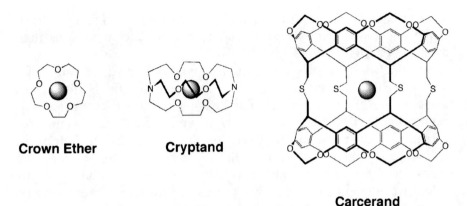

**Crown Ether**    **Cryptand**

**Carcerand**

**Figure 12.2.** Examples of a crown ether, a cryptand, and a carcerand "host" molecule (from left to right). The metal ion or other "guest" molecule is shown as shaded spheres.

tend to be positively charged and the heteroatoms tend to have a negative charge (or partial negative charge) that results in an attractive force (i.e., opposite charges attract, Coulombic forces). In this manner, very stable supramolecules can be formed.

A group of molecules known as the "crown ethers" are the earliest macrocycles to demonstrate host–guest chemistry with metal ions (Figure 12.2). The name crown ether was coined from the zigzag, ring shape of the molecules, which resembles a regal headpiece. The name ether refers to the functional group that makes up the molecule. An ether is essentially an oxygen atom bonded to two carbon atoms; if you don't know this yet, you will by the end of the first quarter of organic chemistry. Like a crown, the crown ether has a central cavity, within which the oxygen atoms hold the metal ion. Charles Pedersen, a chemist at DuPont, achieved the first efficient synthesis of crown ethers in 1967.

Around the same time as the development of the crown ethers, another class of macrocyclic molecules called cryptands also appeared. In this case, the name of the molecule directly derives from its use in host–guest chemistry; a cryptand is a "crypt" for its guest molecule—a darker, but nevertheless accurate description of the host–guest interaction. The French chemist Jean-Marie Lehn of the Université Louis Pasteur first prepared cryptands. Like crown ethers, cryptands are cyclic molecules that can surround and bind metal ions; however, they are generally distinguished from crown ethers by two defining characteristics. First, cryptands generally contain both oxygen and nitrogen heteroatoms. Second, cryptands are not macrocycles, but macro*bi*cycles—that is, they have a three-dimensional, cylindrical shape and hence can more completely encapsulate metal ions than even the crown ethers (Figure 12.2).

Another early host–guest system involved a class of molecules known as the "carcerands." Here again, the name of the molecules derives from their supramolecular host–guest properties, in that carcerand molecules "incarcerate" guest molecules. The inventor of carcerands, chemist Donald Cram of the University of California, Los Angeles, coined the name of these compounds when he reported them in 1985. Carcerands are also three-dimensional, roughly cylindrical-shaped molecules that possess a substantial internal cavity (Figure 12.2). Unlike crown ethers or cryptands, carcerands generally bind small organic molecules as guests, instead of metal ions. This is where we can return to our superhero analogy of the supramolecule. In a study by Cram in 1991, he demonstrated that a carcerand could encapsulate and store a molecule (cyclobutadiene) at room temperature that would otherwise decompose instantaneously. In this sense, the carcerand supramolecule protects the guest molecule from its surroundings and saves the guest from an otherwise rapid demise!

These early examples of supramolecular chemistry—crown ethers, cryptands, and carcerands—are only the tip of the iceberg when it comes to supramolecular chemistry past and present. However, these three examples provide a good basis for understanding the general concept of supramolecular chemistry, supramolecular interactions, and how supramolecular chemistry can be used to discover new science. The three examples also have one other very important feature in common, which is that Pedersen, Lehn, and Cram shared the 1987 Nobel Prize in Chemistry for their work on these molecules. Specifically, the Nobel committee awarded the prize to these three scientists "for their development and use of molecules with structure-specific interactions of high selectivity."

## SUPRAMOLECULES: THE NEXT GENERATION

Now you might ask, what is going on in modern supramolecular chemistry? The answer is both fascinating and complex and depends on exactly what you classify as supramolecular chemistry. In modern chemical research, the lines between different research areas are blurred, such that studies of supramolecular chemistry can overlap with areas as diverse as nanoscience, materials chemistry, chemical biology, and others—drawing well-defined lines between these subjects becomes less meaningful every day. Nevertheless, we can take a look at a few modern examples of supramolecular chemistry that show how the field has progressed and what exciting new directions it might take.

One direction that supramolecular chemistry has taken during the last several decades is the design and study of ever more complex "container" molecules. The earlier versions of such molecules, like those I described above, were only suitable for binding very small guests, such as metal ions, small organic molecules, or, in some cases, small anions (negatively charged ions, like chloride). This allowed for the transport or storage of these small guests but precluded the use of the hosts as a container for doing chemistry. In some of the examples I will describe next, sophisticated host molecules have now been developed that encapsulate multiple, complex guests and hence allow for these guests to interact with each other inside the unique environment provided by the host. Some of the host molecules have been termed "nanoflasks" or "nanoreactors," that is, supramolecules within which other molecules can be placed so that they undergo chemical reactions inside the host. Often, the outcome of the reactions of guest molecules inside one of these hosts will be different from what occurs if the guest molecules interact with each other outside (or in the absence of) the host molecule. As you will read, chemists are pushing supramolecular chemistry closer to the scale and complexity of the viral capsid. We can now make and study molecules that can create completely new chemical environments within which new chemical reactions may be discovered!

Many of the distinctions between different host molecules, or nanoflasks, arise from the different weak intermolecular forces that are used to put them together. Many of the most widely studied nanoflasks use either hydrogen bonding or metal ion binding to assemble themselves. I will tell you about a few examples that encompass both types of interactions, which should give you an overall sense of these types of nanoflasks.

Nanoflasks based on hydrogen bonding have been studied extensively for the last several decades. You may recall that hydrogen bonding is a weak electrostatic interaction that occurs between a hydrogen atom, which acquires a partial positive charge when bound to a heteroatom (e.g., oxygen or nitrogen), and a second heteroatom that possesses a partial negative charge (remember, opposite charges attract). Although weak on an individual basis, when combined in large numbers or carefully organized, hydrogen bonding can lead to a cumulatively large interaction between molecules. An everyday example of this can be found by comparing the boiling point of water with that of many hydrocarbons, like propane. Even though propane has a molecular weight 2.5 times greater than water, it has a boiling point of $-42°C$, far lower than that of water. This is due, in large part, to the strong hydrogen bonding that occurs between water molecules, making water exceptionally stable as a liquid despite its very small size.

The coordinated arrangement of multiple hydrogen bonds has been used to generate a variety of molecular containers that display a wide array of fantastic chemical traits. Many groups around the world have studied these supramolecules, so I will just highlight the work of one of our San Diego neighbors. Professor Jules Rebek of The Scripps Research Institute (TSRI), just down the road on Genesse Avenue, is one of the world's leading scientists in the area of self-assembled nanoflasks. Professor Rebek's research has focused on molecules that assemble together via hydrogen bonding. He has made a variety of important contributions to the field of supramolecular chemistry, but among his most famous is his molecular "tennis ball."

Rebek devised the name molecular tennis ball to describe a supramolecular host of two curved molecules that snap together along a "seam" of hydrogen bonds in a way that is reminiscent of a tennis ball (Figure 12.3). The tennis ball is stable both as a solid and in solution and can trap organic guest molecules such as methane and ethane inside its cavity. Rebek and his colleagues showed that the basic idea behind the tennis ball could be easily generalized to make molecular balls of many sizes. As you might expect, the molecular "hacky sack ball" is a smaller version of the tennis ball that can hold methane as guest, but nothing larger. In contrast, the molecular "softball" is significantly larger than the tennis ball and can accommodate large guests, such as tetramethyladamantane. Perhaps more importantly, the softball can also accommodate more than one smaller guest molecule, and this opens the possibility of the nanoflask—using the softball as a tiny vessel within which unique chemistry can be pursued.

Indeed, Rebek has shown that the molecular softball can encapsulate two guest molecules and that a suitable combination of guests can undergo a reaction with one another within the softball. For example, it has been shown that cyclohexadiene and *p*-benzoquinone (Figure 12.3), two different guest molecules, can enter the softball and undergo a Diels–Alder reaction (again, if you don't already know

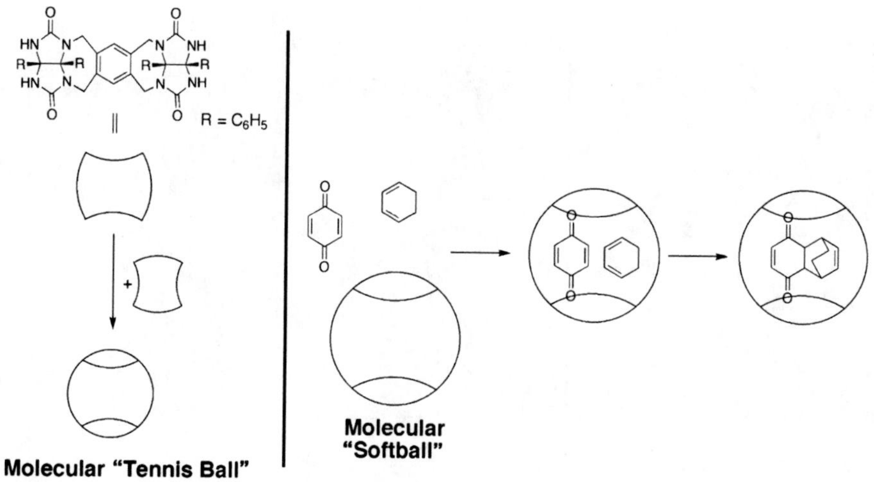

**Figure 12.3.** The molecular tennis ball is made up of two identical hydrogen-bonding molecules that have a concave shape (left). A similarly shaped molecular softball (made up of similar but larger concave hydrogen-bonding pieces) can catalyze the Diels–Alder reaction between cyclohexadiene and p-benzoquinone (right).

the Diels–Alder reaction, you will learn about it in organic chemistry this year). Not only do the molecules undergo this reaction, but they do so 180 times faster than when they are outside the softball. In other words, the softball can act as a catalyst by confining molecules within its cavity and by accelerating the rate of reaction. Here we see how supramolecular chemistry can lead to the discovery of unique chemistry that is not observed under "normal" circumstances.

In addition to hydrogen-bonded host molecules, like the hacky sacks, tennis balls, and softballs described above, metal ions have been widely used to stitch together nanoflask capsules. In contrast to early supramolecular systems, where metal ions are the *guest* species (e.g., crown ethers, cryptands), in many of the newer systems, the metal ions are a key part of the *host* molecule. Many people have contributed to this area of metal-driven supramolecular capsules, and one of them was my PhD advisor, Professor Kenneth N. Raymond. When I started my PhD at the University of California, Berkeley, in 1994, Professor Raymond was just starting to work in this area. I did not personally work on this subject while I was earning my PhD, but I was witness to its development in Professor Raymond's laboratory, and it gave me a great perspective on the subject.

When Professor Raymond and his team became interested in metal-driven supramolecular capsules, they nicknamed the project the "Borg" project. This reference will be totally lost on you unless

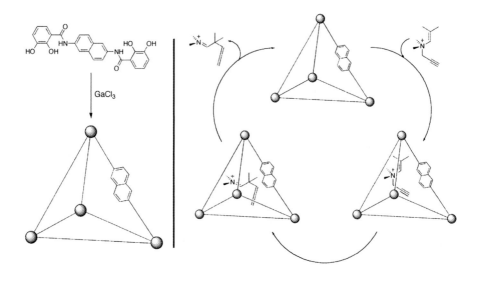

**Figure 12.4.** *A tetrahedral nanoflask is prepared from a bis(catechol)dinapthalene ligand and a gallium(III) salt (left: the ligand is shown as lines and the gallium ions as spheres). This flask acts as an artificial enzyme, catalyzing the aza-Cope rearrangement of organic cations within the nanoflask interior.*

you are a fan of *Star Trek*—I will assume you are not. In that case, the Borg were a malevolent alien species on Star Trek that were cyborgs—part living being, part machine—or as Professor Raymond's team looked at it, part organic, part inorganic (Figure 12.4). In addition, the Borg's spaceship was a giant, flying cube, which was the shape of one of the molecular capsules that Raymond and his coworkers wanted to create. So, as you can see, the nickname was quite apt—the synthesis of highly symmetric molecular capsules (e.g., cubes and tetrahedra) made of organic ligands and inorganic metal ion components. That's enough science fiction; let me tell you about science reality.

Professor Raymond's team has successfully developed a number of capsules since 1994, but one system, based on a bis(catechol)dinapthalene ligand and trivalent metal ions such as iron(III) or gallium(III), has been the most interesting. The symmetry of the organic ligand (twofold) with the preferred symmetry of the metal ions (threefold) drives the combination of the two (ligand and metal) to form a molecular capsule with the shape and symmetry of a tetrahedron

(Figure 12.4). Like the other capsule molecules I have already described, this molecular tetrahedron was able to accommodate a wide variety of guest molecules within its hollow core. In this case, positively charged molecules (cations) are preferentially taken up into the tetrahedron because the host carries an overall net negative charge (due to a charge mismatch of the metal ions and catechol ligands). In addition to encapsulating different cations, the tetrahedron, like some of the other supramolecular assemblies described above, was able to stabilize certain species and catalyze certain reactions. For example, in a very recent report, the tetrahedron was shown to encapsulate the propargyl enammonium cation that can then undergo the aza-Cope rearrangement (again, you will learn this in organic chemistry). The rearrangement reaction is accelerated as much as 184-fold inside the tetrahedron (Figure 12.4). In addition, after rearrangement, the iminium ion product is released from the tetrahedron, allowing for the process to continue. So, the tetrahedron not only acts as a catalyst but also behaves like an *artificial enzyme*, bringing the concept of nanoflasks and the world of supramolecular chemistry a step closer to mimicking the magic of nature.

As I mentioned earlier, a number of researchers have contributed to this area of metal-driven capsules, and if you want to learn more, you can look up some of the papers published by Professor Raymond. Another body of work you might want to look at is that of Professor Makoto Fujita of the University of Tokyo, Japan. Professor Fujita was among the early pioneers in this area and has made some of the largest, most complex, and most amazing molecular capsules ever described. Recently, his research group described large, spherical capsules he refers to as "nanodroplets." These nanodroplets are metal-assembled capsules that can be designed in such a way as to create an artificial, liquidlike environment within their interiors. Professor Fujita's work is another wonderful example of how the chemistry is starting to catch up with biology.

## FROM MOLECULES TO MOFS

OK, thus far, I have given you a modest history of supramolecular chemistry and some of its more recent incarnations. Before I lose your attention, let me finally discuss some of the supramolecular chemistry that we are doing in my research group. I hope it will spark your imagination a bit.

All of the supramolecular chemistry I have told you about so far involves detached, isolated supramolecules that involve a host–guest relationship. To understand the research in my laboratory, we have to think in terms of more extended molecular structures; that is, we are going to make our host molecule not just a single capsule molecule but rather a series of capsule molecules, congregated together in a continuing lattice. When we connect enough of these capsules together, we get a solid material composed of many individual, identical chambers. This is the realm I work in, the realm of MOFs.

MOFs are materials composed of organic molecules and metal ions; in principle, they are formed using the same design concepts developed for the metal-driven capsules of Professor Raymond and Professor Fujita. However, in the case of MOFs, the organic molecules and metal ions are selected so that an infinite lattice of chambers is formed, instead of individual, separate supramolecular capsules (Figure 12.5). Jungle gyms or the scaffolds of buildings have been used as analogies for MOFs, where a structure is composed of rods (organic molecules) connected by nodes (metal ions). Like the scaffold of a skyscraper, an MOF is a structure that occupies a large amount of space, most of which is unoccupied (hollow). Think about it—if you have the steel scaffold of a 20-story skyscraper assembled, the overall structure is the size of a 20-story skyscraper (duh!), but

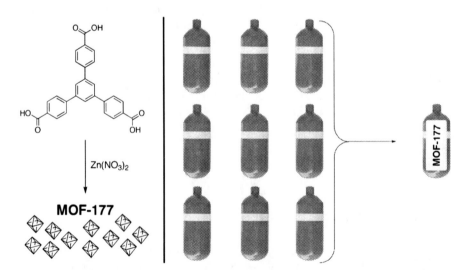

**Figure 12.5.** *Combining 1,3,5-benzenetribenzoic acid and a zinc(II) salt gives the metal–organic framework MOF-177 (represented by polyhedral crystals). Filling a tank with MOF-177 allows it to hold as much carbon dioxide gas as nine empty tanks (right). This highlights the potential importance of MOFs as gas storage materials.*

most of the space inside the structure is empty because the beams and rivets only occupy a small part of the overall volume. It is the same for MOFs, where the lattice, composed of organic molecules and metal ions, is sturdy and rigid, and yet most of the space within the MOF is empty. The empty space in the skyscraper scaffold will eventually be filled with the floors, walls, and so on, followed by furniture, computers, and people. Similarly, the empty space in the MOF can be filled by many different types of guest molecules, which ultimately results in interesting chemistry and interesting, new supramolecular materials.

The term MOF was coined by Professor Omar Yaghi of the University of California, Los Angeles. These types of supramolecular materials have also been called coordination networks or porous coordination polymers (PCPs). The use of the term "coordination" refers to the type of bond between the organic molecules and the metal ions, which are often referred to as coordination bonds. Materials that fit the description of MOFs were described prior to Professor Yaghi's coining of that particular term, but since then, the use of the term MOF has become the dominant acronym for these materials.

Many different types of MOFs have been produced in the last decade or so. Among the most widely studied MOFs are those described by Professor Yaghi, which are composed of rigid di- or tricarboxylic acid molecules and metal ion clusters of zinc (Figure 12.5). Again, the organic, carboxylic acid molecules act as rods, and the metal ion cluster acts as a node or linker. In cases where the node is not made up of a single metal ion but rather a cluster of metal ions (as in the case of most of Professor Yaghi's MOFs), the term secondary building unit (SBU) is often used to describe the node. When the organic acid molecules and zinc ions are combined and heated in solution together, MOFs will form and appear as crystals (like the diamonds in your mother's wedding ring) out of the solution. Many crystals of MOFs are incredibly stable, able to withstand temperatures in excess of 400°C. Using a variety of different organic molecules and metal ions, a vast array of MOFs has been synthesized and studied.

You may be asking yourself, why are MOFs being studied? What are they good for? Well, like many areas of science, particularly younger ones like supramolecular chemistry and MOFs, the answer is we don't know yet. However, MOFs do have some interesting host–guest chemistry that is making for some facinating research and is even getting closer to some very important real-world applications.

In recent years, the main interest in MOFs has been in the area of storing gases. Why gas storage? Because all of that empty space does a very good job of storing large amounts of gases, such as hydrogen, methane, and carbon dioxide (depending on the MOF). Gases like hydrogen and methane are important as fuels (think hydrogen fuel cells and methane-powered cars), and carbon dioxide is an important greenhouse gas that needs to be removed from the atmosphere and somehow stored and/or immobilized. So, with respect to these technologies, the storage of the aforementioned gases is a very important scientific problem.

With respect to gas storage, MOFs are fantastic hosts for gaseous guests—indeed they are the best ever known. Among the greatest success stories to date is a material known as MOF-177. MOF-177 can perform some amazing feats of gas storage. Because of its huge amount of void space, and the fact that MOF-177 can weakly interact with gases, MOF-177 can actually improve the storage of gases. This is best illustrated by the following experiment: a pressurized gas cylinder (at room temperature and a pressure of 30 bar) can hold a certain amount of carbon dioxide. Take that same empty tank and fill it with MOF-177. Now put the carbon dioxide gas into the tank full of MOF-177. Here's the question: does the empty tank or the tank full of crystals of MOF-177 hold more gas? Instinctively, you might think that the empty tank will hold more gas because in the other tank, the solid MOF material is taking up space. However, exactly the opposite is true (Figure 12.5); the tank full of MOF-177 holds almost nine times as much carbon dioxide as the empty tank! Stop and think about that—it is pretty amazing that a tank full of a solid, crystalline material (MOF-177 looks somewhat like table salt) holds more gas than an empty tank. Again, the reason for this is the high surface area of the MOF, which gives more places for the gas to absorb. And when we are talking high surface area, we are talking really high, ~4500 m$^2$/g, which translates to 1 tsp of IRMOF-1 having the surface of several football fields! So, now you might start realizing the potential importance of MOFs. Take a tank with MOF-177 and you can safely increase the capacity of storing greenhouse gases such as carbon dioxide. This has the possibility of having significant impacts on the energy, environmental, and transportation sectors in the future. Indeed, the U.S. Department of Energy (among other agencies) has taken a great interest in, and invested significant amounts of government research dollars in, learning more about MOFs and how they might contribute to our current energy concerns.

## "INTERIOR DECORATING" WITH MOFS

MOFs are highly porous materials, with large surface areas that might be interesting for gas storage. So what are we doing in my laboratory? Well, we have worked on a few topics, but most recently, we have focused on something we are calling the postsynthetic modification of MOFs. "What the heck is that?!" you ask. Well, I will tell you; it is really rather simple.

By performing a postsynthetic modification of MOFs, we are becoming the "interior designers" of MOFs. So, let's go back to our empty building analogy. So we have the framework for a building; let's put in some floors and walls, with doors and windows connecting the rooms; the building is still mostly empty space. So, as interior designers, we come in and hang some paintings, put in some furniture, add a desk; that is, we make the interior of the building functional. We take a fully constructed building, which starts as a series of empty rooms, and we add components to each room to make them useful for some purpose. In the postsynthetic modification of MOFs, we do the same thing; we build an MOF, which, once formed, is a series of empty, interconnected chambers, and then we add a reagent that flows through the MOF (remember it is a very open structure) and modifies each chamber, and in the process making each chamber more "functional" than it was before! We use the term postsynthetic modification to refer to the fact that we functionalize the MOF once it is already completely assembled. This makes sense in our analogy; would you hang pictures on a wall and *then* install the wall in a building? If you tried to, the picture could fall off, break, and so on, so you put the walls up first, then the decorations. We do the same thing in postsynthetic modification: we build the "walls" of the MOF first, and then we hang our more delicate functional groups postsynthetically (*after* the MOF has been assembled).

Recently, we have found postsynthetic modification to be a really useful approach to making highly "functionalized" MOFs. To do this, we have largely worked with IRMOF-3, which was first described in Professor Yaghi's laboratory. IRMOF-3 has amino groups on the organic rods, which point into the cavities of the MOF. From a chemistry perspective, these amino groups are like hooks that are already in place on the walls of our building, so all we need to do is go in and hang something on them. By using reagents that are reactive toward amino groups, we have found that we can indeed

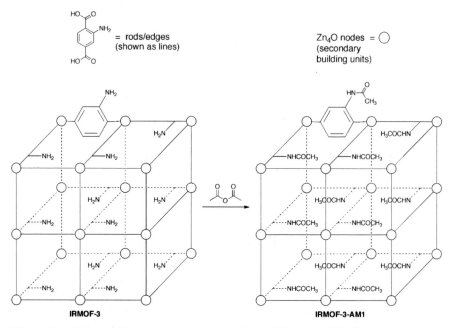

**Figure 12.6.** Diagram illustrating the postsynthetic modification of IRMOF-3 (shown as the cubic lattice) with acetic anhydride to generate IRMOF-3-AM1. Some of the amine groups are not shown for clarity. One of the ligands (rods) is shown more explicitly for illustrative purposes.

decorate every chamber in the MOF. In one specific example, we have used acetic anhydride to convert all the amino groups in IRMOF-3 to amide groups (Figure 12.6). We have many other examples showing that we can modify different MOFs with many different types of reagents. We are hoping that this new method of modifying the insides of MOFs can be used to enhance their properties, including the absorption of gases like methane and hydrogen.

Anyhow, Angela, that is probably all that you want to read about supramolecular chemistry for now. I hope you have found my letter interesting and are not drooling all over the paper because I have put you to sleep. If you have any questions or interest in my research group, feel free to stop by my office in Pacific Hall. I hope you will choose to do some undergraduate research, if not in my own laboratory, then in some other—there are a lot of great research opportunities here at the University of California, San Diego. In the meantime, have a great sophomore year and keeping doing well in your chemistry classes!

Sincerely,

Seth Cohen

## FURTHER READING

Cram, D. J. The design of molecular hosts, guests, and their complexes (Nobel lecture). *Angewandte Chemie International Edition in English* 1988, *27*, 1009–1020.

Eddaoudi, M.; Moler, D. B.; Li, H.; Chen, B.; Reineke, T. M.; O'Keeffe, M.; Yaghi, O. M. Modular chemistry: Secondary building units as a basis for the design of highly porous and robust metal–organic carboxylate frameworks. *Accounts of Chemical Research* 2001, *34*, 319–330.

Fiedler, D.; Leung, D. H.; Bergman, R. G.; Raymond, K. N. Selective molecular recognition, C–H bond activation, and catalysis in nanoscale reaction vessels. *Accounts of Chemical Research* 2005, *38*, 351–360.

Rebek, J. Jr. Reversible encapsulation and its consequences in solution. *Accounts of Chemical Research* 1999, *32*, 278–286.

# 13

## Biomaterials at the Beach: How Marine Biology Uses Chemistry to Make Materials

**Jonathan J. Wilker**

*Purdue University*

Jonathan Wilker is an associate professor of chemistry and materials engineering at Purdue University in West Lafayette, Indiana. His research program focuses on materials produced by marine organisms. Ongoing efforts include characterization of marine biological materials, developing synthetic polymer mimics, and designing applications for these new materials. Projects are often inspired by what is seen while scuba diving.

Hi Angela,

Put down the books! (OK, not this one.) We're going to the beach!

**Figure 13.1.** Some sticky sea species that you will find in tide pools: mussels (left), kelp (center), and barnacles covering a rock (right).

## YOUNG SCIENTIST, GO SEAWARD

Let's go tide pooling and look for interesting sea creatures. Pretty much any rocky coastline spot will do. We're timing our arrival for around low tide. You happen to live close by to one of my favorite spots—Laguna Beach in Southern California. When driving toward the coast on Route 133, you will end right at the crescent-shaped beach. Look to your right and there is a huge, rocky outcrop—tons of cool things we can explore there. Wear sandals or old sneakers that you can get wet, because the rocks, shells, and coral are pretty sharp. Be careful not to slip on the seaweed.

While you're crawling over the rocks, you can see tons of mussels, barnacles, sea urchins, anemones, seaweed, coral, starfish, periwinkles (snails), and tiny fish. Some of these critters are shown in Figure 13.1. If you get lucky, you might even spy a lobster or a crab hiding in a crevice. Once in a while, pick up your head and look out to sea. This spot is good for sighting the occasional sea lion barking at you or even dolphins and gray whales a little farther out.

It may not be obvious right away, but lots and lots of fascinating chemistry is going on here under your feet. Be careful not to crush too much of it! Maybe best of all is that most of the chemistry is unknown and just waiting to be explored.

Grab a hold of a mussel or two. Try pulling it off the rock. Pretty difficult, isn't it? What about one of those tubes made by the marine worms? You'll need a hammer and chisel to get at them. And those barnacles? Don't even bother trying to pull them off the rocks. No chance. Soooo … how do you think these creatures are able to attach themselves so well? Some quick observation on your part will tell you that they are making adhesives and cements. OK, so what about the chemistry of these glues? Interestingly enough, we don't really know yet. We do have some information, but, for the most part, the chemistry of these materials has not yet been figured out. Look at any of the little animals around you, there in the tide pool. Ask yourself a simple question, like maybe about how a marine biomaterial is made. Or you might wonder what, exactly, a cement is composed of. Or how it functions. Or what types of chemical bonding are going on within. Seemingly simple stuff to ask, really. But odds are that the answers to your questions are unknown. To me, that's where the most exciting science is—where so very little is known.

## CHARACTERIZING MARINE BIOLOGICAL MATERIALS

What all this means to a budding chemist is that tons of fascinating chemistry is out there, just waiting for people like you to start exploring it. You have an open research field in which to roam about, to make new discoveries, and to design new technologies. It's all far too interesting to pass up. So what do you want to know? Well, for starters, let's try and figure out how nature makes materials. We can talk about hard materials like seashells or soft materials like mussel adhesive. We want to identify the key types of molecules from which the materials are made. We want to learn, say, how a biological adhesive sticks to a rock. We can ask what is the mechanical performance of these materials. Or perhaps we may wonder what surface these animals will *not* stick to.

You may start asking your questions from the perspective of a chemist. But one of the many fun aspects of this type of research is that you will soon also be thinking about issues in biochemistry, marine biology, materials science, and materials engineering. You have lots of opportunities to be creative here. There are many more questions than you will ever have time to answer. So follow your nose to whatever you find to be the most intriguing.

We always have a pretty wide variety of studies going on at any one time. We may start with field work to observe the sea creatures and how they function, then we may collect some of them to bring back to the lab. We have built large aquarium systems in our lab for growing and observing the animals. While the creatures are growing in the tanks, they also produce their sticky materials for us to collect and study. We can assess the mechanical properties of the materials. For example, we may convince mussels to deposit their adhesive onto different substrates and then measure how strongly they adhere when we change the surface or water chemistry. Then, we can use a host of chemical spectroscopy techniques to start characterizing what the glue is. Insights on the types of molecules within the glue or the kinds of bonding motifs present may arise.

Another way to attack the characterization of biological materials is to try and extract out molecules. If we can render at least part of the material soluble, it might be a little easier to figure out what we have. If we can pull, say, a protein out from the solid, then we are able to engage in a host of biochemical studies that will help us understand what is going on. Small molecules may be found in there as well. Sometimes, we go about synthesizing small molecules to try and mimic what the larger, intact material may look like. If we find some function or similar reactivity in these small molecules, it helps to show what kind of chemistry could be going on underwater. Another fun thing we do is to synthesize large molecules that mimic the chemistry we discover in the sea animals. Then, we combine these large molecules into functional materials to mimic what we get from the ocean. So, you may start by characterizing a marine bioadhesive and end up with a new synthetic glue that works underwater.

Overall, I think it's all loads of fun. We are working to reveal how nature makes materials. Very little is currently known so that we're bound to discover something new all the time. On its own, that's enough to keep us excited while working in this field. But there's a lot more to it than that. We can easily imagine that knowledge about these materials will translate into development of various applications.

Wouldn't it be great if, after a surgery or injury, you could glue your skin or internal tissues back together? Just avoid the pain and additional damage from sutures. There aren't any good glues to do that right now. Most glues you've seen need to dry out before they stick. But that definitely won't work with an internal biological environment where it's always wet. Marine biology has answers for us. After all, these organisms are adhering underwater. And how about trying to prevent all these marine organisms from sticking to ship hulls? You might figure out how they stick and then develop surfaces to repel them. Then ships could travel through the water more easily, burning less fuel.

You can see that there are plenty of applications that also arise from this line of chemistry research. In the following pages, we will discuss what is known about how marine biology makes materials and how they function. Then, we will look at how people are trying to make their own versions of these materials. Specific applications for sticking, and for preventing sticking, will also be shown. So keep your feet wet. We have lots of chemical exploring to do!

# HOW DO THEY STICK?

Looking all around you in the tide pools, you can see so many different creatures sticking themselves to the rocks—tube worms, acorn barnacles, gooseneck barnacles, limpets, mussels, seaweed, periwinkles, starfish, and anemones (see Figure 13.1). We don't yet know how most of them stick. So there is a lot of fun science for you to explore in the future. For now, let me give you an overview of what we *do* know about these critters. We scientists have more information on mussel adhesive than any other marine system.

## Mussel Adhesion

From time to time, we go out into the field and collect mussels from beaches throughout New England. We bring them back to our lab for growth in tanks. Figure 13.2 (left) shows a mussel sticking to a sheet of glass. You can see that the adhesive system is composed of threads emanating out from the shell. Each thread is terminated with a small disk (or "plaque") of adhesive. That's the glue actually in contact with the glass surface. These adhesive plaques are pretty small—only about 2 mm across. While at the beach, the adhesive is usually covered up by the shell and is a little difficult to see. But if you pull a mussel off and look closely, you can see the glue. Figure 13.2 also shows a

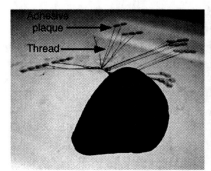

**Figure 13.2.** Mussels sticking to a sheet of glass (left) and Teflon (right).

mussel we placed onto a sheet of Teflon (polytetrafluoroethylene). He's sticking there just fine. Now that's a pretty impressive glue, sticking to Teflon like that.

Those adhesive plaques are a protein-based material. Proteins are long, large molecules. In many ways, proteins are simple polymers— lengthy molecules made up of many linked monomers. In the case of proteins, there are typically 20 different monomeric units, known as amino acids. Proteins can often be composed of more than 500 amino acids connected together and have molecular weights over 100,000 mass units.

When a mussel wants to stick to a surface, the animal excretes a mixture of proteins. Then the proteins are cross-liked to cure the material. Although not a super common process, protein cross-linking is seen from time to time, more often in hardened materials. Cross-linking is a process by which the various long protein molecules are linked to each other with chemical bonds, depicted in Figure 13.3. This cross-linking helps to lock the whole system in place. Think about the movie *Raiders of the Lost Ark* when Indiana Jones is dropped into that pit of writhing snakes. All those snakes slithering make it look as though the whole floor is moving around. Imagine if you could use a rope to tie each snake to one or two of its neighbors. The whole mass of snakes would then not be able to wriggle around any more. Protein cross-linking is, in many ways, much like tying all the snakes together. In the case of mussels, once cross-linked, the protein-based glue is cured and locked down on the surface.

So you might wonder how, exactly, mussels make chemical bonds to cross-link one protein to another. The reactivity starts with a rather unusual molecule in the adhesive proteins called 3,4-dihydroxyphenylalanine (DOPA), shown in Figure 13.4 (left). As a small molecule on its own, DOPA can be a drug used to combat Parkinson's disease. When bound within a protein chain (Figure 13.4, right), mussels take advantage of DOPA's reactivity to bring about cross-linking. Seeing DOPA in a protein is quite rare, and nature is

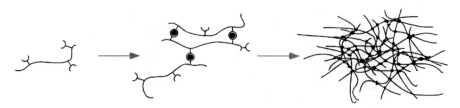

**Figure 13.3.** *A general schematic of cross-linking long protein molecules.*

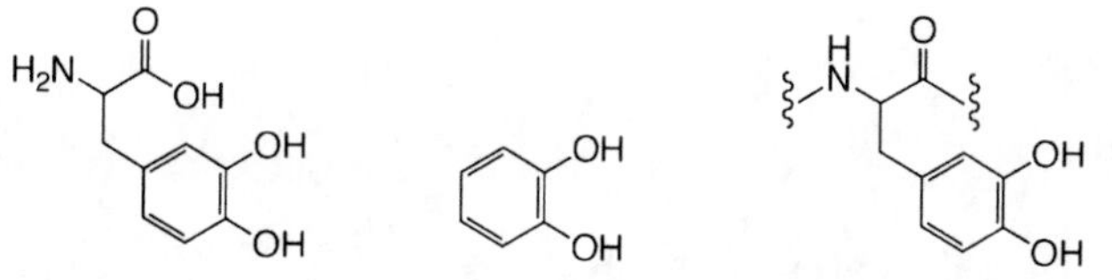

**Figure 13.4.** *A single molecule of DOPA, catechol, and DOPA bound in a protein chain.*

usually very efficient—if something is there, it is there for a good reason. DOPA can be oxidized more easily than any of the other standard 20 amino acids used in biology to make up proteins. DOPA also has the ability to bind (or "chelate") metal ions quite well. It turns out that this combination of abilities to bind metals and be oxidized easily is likely central to the cross-linking of mussel proteins and the formation of their adhesive.

We have been working with live mussels, extracted proteins, and smaller synthetic peptides (or pieces of proteins) to try and determine the reactive chemistry at the heart of mussel glue formation. You wear many hats in such studies. One day, you might be a synthetic chemist and another day a marine biologist, a biochemist, or a materials engineer. Each class of experiments provides a different perspective on a research problem. If things are going well, you can get complementary data where all the experiments point to the same conclusion. In the case of mussel adhesive, we have come across some pretty interesting chemistry.

Mussels take iron from the seawater around them. Although there is not a very high concentration of iron (or other metals) in seawater (e.g., 1–10 parts per billion), there is plenty to sustain all the life that you see when at the beach. Mussels collect iron from the surrounding seawater and shoot this iron into their adhesive. The resulting glue has about 100,000 times more iron than the water in which mussels live. Pretty interesting, huh?

Through a series of different studies, we found that iron in the glue is bound by DOPA groups of the adhesive proteins. More specifically, three DOPAs bind each iron to generate the metal–protein complex shown in Figure 13.5. In particular, this iron starts out bound to the DOPAs in the $Fe^{3+}$ oxidation state. Oxygen ($O_2$) then reacts with the Fe(DOPA)$_3$ complex and generates a radical species in the protein (Figure 13.6). The iron is reduced ($Fe^{3+} \rightarrow Fe^{2+}$), and the DOPA group is oxidized at the same time. Oxidation of DOPA is removal of an electron (given to the $Fe^{3+}$) to yield the radical.

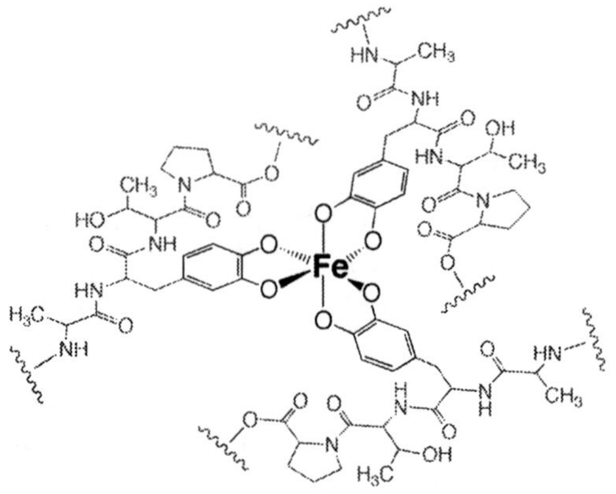

**Figure 13.5.** The Fe(DOPA)₃ complex proposed to be central to protein cross-linking in mussel adhesive.

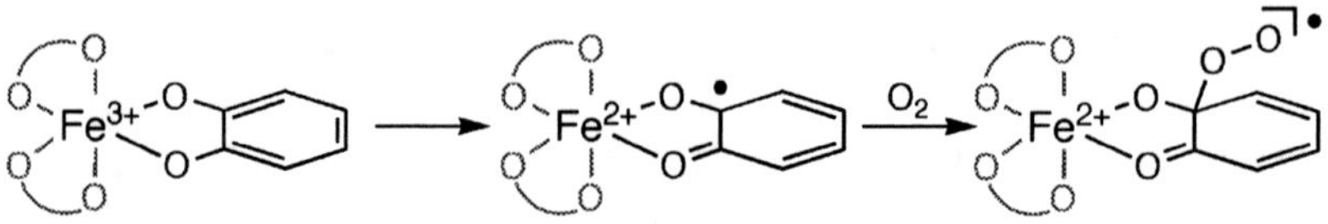

**Figure 13.6.** Oxidation and reduction in mussel adhesive to generate a reactive radical. The black dot represents the radical. The DOPA ligands bound to the Fe center, seen in full in Figure 13.5, are simplified here and show only one ring.

Of all the types of molecules you will see in chemistry, radicals are the most reactive. So when you form a radical, things happen fast and indiscriminately. At this point in the process, we have less in the way of detailed chemical insights. But our current suspicion is that these radicals can do two things with regard to forming the mussel glue, both of which are shown in Figure 13.7. First, the radicals are likely to react with one another, thereby generating covalent, organic bonds between proteins. Here we get protein–protein cross-links to cure the material. These cross-links are like the snakes being tied together with rope. This first process is often called "cohesive bonding," in which the material interacts with itself.

Second, the radicals may be able to react directly with the surface to which the mussel is trying to adhere, thereby creating covalent bonds to the surface. This second process is called "adhesive bonding," in which the glue actually bonds to the surface. You need both cohesive and adhesive bonding to make a good glue. If you only have cohesive bonding, your glue bonds to itself, making a hardened ball, but ignores the surface. If you only have adhesive bonding, you will

Cohesive protein cross-linking:

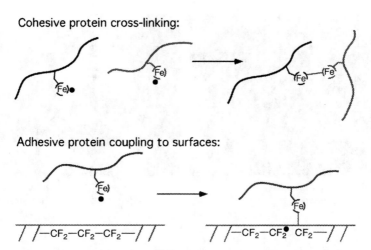

Adhesive protein coupling to surfaces:

**Figure 13.7.** *Schematic of radical reactivity in mussel adhesive. Protein–protein cross-links and protein–surface bonding are the results. Note that the radicals shown are simplifications of those at the metal center, seen in Figure 13.6.*

make a thin layer of material on the surface, but the rest of the glue will not stay in place.

Put all of our chemical insights together and we now have a picture of how mussels cross-link their proteins. Some of the details need to be figured out, but we are still working at it and are getting there. You can see that these mollusks are quite diverse in that they combine inorganic chemistry, organic chemistry, protein biochemistry, and materials science to make their intriguing glue. What about all the other creatures in the sea? How do they make adhesives and other materials? Far less is known, so there is a lot of room for you to jump in and characterize marine biomaterials with us. Let's take a look and summarize what we do know at this point.

## Other Marine Biological Materials

***Barnacles.*** After mussels, we probably know the most about how the common acorn barnacle makes its cement. You can see barnacles on a rock in Figure 13.8. Here, too, the material is based on cross-linked proteins. However, there is no DOPA in the barnacle cement proteins. Perhaps the thiol groups of cysteine amino acids (i.e., RSH groups) are oxidizable to disulfides and take part in an oxidative protein cross-linking to form barnacle cement, although we do not yet know for sure. Hair perms rely on just this type of disulfide formation to link hair and hold it in a given configuration. Barnacle

**Figure 13.8.** Barnacles on a rock (left) and an underwater photograph of tube worms on a rock (right).

cement appears to be composed of about ~90% protein versus mussel adhesive, which is at least 99% protein. The remainder of barnacle cement is about 5% inorganic and has small fractions of carbohydrates and lipids as well. Some intriguing, recent results indicate that there may be unexpected analogies between the chemistry of barnacle cement curing and human blood clotting.

**Tube Worms.** The lowly tube worm does some pretty impressive materials engineering. In Figure 13.8, you can see the tubes created by polychaete or Christmas tree worms. It was pretty murky the day I took this underwater photograph, but you may still be able to see the tubes and dark red worm heads, in the left of the frame, resembling a Christmas tree. You can find worm tubes such as these all over rocky beaches in Southern California. These guys synthesize a glue, but only a little bit of it. While they are making this glue, they also collect grains of sand, bits of broken shell, and other solids from their surroundings. Then they combine their organic glue with these small inorganic particles to fabricate a true composite material. Very efficient use of materials! With regard to specifics of the glue, we do have a few insights. The glue is protein-based and does contain some DOPA. These glue proteins tend to be anionic (i.e., negatively charged), and the worms might be mixing their anionic glue proteins with cations such as $Mg^{2+}$ and $Ca^{2+}$ for condensation of the proteins and formation of the final material.

**Limpets.** In Figure 13.9, you can see a limpet, a pretty common sight on most rocky beaches. Limpets use both glue and a big suction cup for sticking to surfaces in the intertidal zone. At low tide, when these

*Figure 13.9.* A limpet on a rock (left), a sea cucumber underwater (center), and the bottom of a starfish, seen through a sheet of glass (right).

shellfish are exposed, they tend to stay in place and rely on an adhesive for attachment. At high tide, the animals are underwater and move around to forage. Then they are using their big suction cup to stay in place or to release and move. The glue is not terribly well characterized, but we know that there is protein in the adhesive and that there is about a 50% inorganic component to the material as well.

**Sea Cucumbers.** You probably won't see too many sea cucumbers in tide pools, but go scuba diving in the right spots and you will find plenty (Figure 13.9). These holothurians have a rather unique defense mechanism. When they are threatened by a predator, the cucumbers eject a very sticky, quick-setting glue to tangle up their pursuer. While the thwarted predator is tying to untangle itself from the mess, the sea cucumber runs away to safety. Here, too, we do not have many details on the workings of this bioadhesive, but we know that the glue is about 60% protein and about 40% polysaccharide in composition.

**Starfish (or Sea Stars).** Looking around the tide pool, you're bound to see lots of starfish. Give them a close inspection, or see Figure 13.9, in which we have a starfish on a glass sheet and are looking at the animal's bottom. See all the little tiny tendrils (i.e., "podia") the starfish uses to move around? Each of these podia is tipped with an adhesive material. Starfish walk around by stretching out some podia, sticking to the ground, pulling the rest of the animal along with them, detaching the podia, and repeating this process many times. Wait around at the tide pool long enough and you will spy a starfish eating a mussel. It's a pretty amazing thing to watch. The starfish will wrap itself all around the mussel, glue itself to the mussel's shell with the podia, and then tear open the shellfish so it can get at the meal inside. One interesting question is what glue do starfish use to stick? Another question is how do they *un*stick? At this point, about all we can really say is that their adhesive contains proteins, polysaccharides, and lipids, and that the inorganic content is about 50% of the total material.

You can see that, so far, we have uncovered some pretty interesting chemistry to explain how sea animals make their adhesive materials. But there is far more unknown than known. Lots of opportunities await new researchers, like yourself, to join the field. Do you want to devise a new research project? Just look around you in the tide pool. You can surely find a few animals that make materials with interesting properties. I bet that the chemistry used by most of them is completely unknown.

## HOW CAN I GET MY HANDS ON SOME? MILKING MUSSELS FOR GLUE

Now that you've been to the beach and tried pulling some of these sea creatures off the rocks, you are likely pretty impressed. You can easily imagine that there might be lots of cool things to do with these glues, such as develop surgical adhesives or dental cements. But that's only going to work if you can get a lot of the material. I've got news for you: That's not easy. Remember the mussel glue from Figure 13.2. Those adhesive plaques are tiny and you can't get any useful glue from the plaques because it's already cured. If you want to extract mussel glue, you need to get the proteins from inside the animal before they have been deposited onto a rock. We have done this in our lab far too many times to even count. The reason we have had to extract protein so many times is that you only get minuscule amounts of adhesive protein from each extraction. Maybe 20 mg each shot, at most. In order to get a mere gram of mussel adhesive protein, you would need to start with about 10,000 mussels! With the work involved, it would take one person two solid months in the lab to extract that single gram of glue. For real applications, we will be needing *kilo*grams of the glue, if not tons. Nope, extracting the real stuff is just not practical at all. We have to be clever and look at other ways to get our hands on the bioadhesives.

Biotechnology and genetic engineering provide one potential route. We can engineer bacteria, such as the common *Escherichia coli* (or *E. coli*) to actually produce proteins that they normally do not have. In this case, we might convince a bacterium to start cranking out mussel or barnacle adhesive proteins. This approach does have some promise for the future, and many people are looking at it closely. You may have heard the case about a company that developed genetically engineered goats to produce spider silk, another interesting protein-based biological material. The goats produced the silk proteins in their milk, and in decent quantities at that. But one of the problems with this approach turned out to be that the goats lack the complex spinneret system that spiders have. So, while the goats could make the proteins, the final processing was not there and this genetically engineered silk protein never quite had the same properties of silk from real spiders. That company went out of business.

Synthetic chemistry has progressed by leaps and bounds in recent decades, especially with regard to the ability to make peptides and even small proteins. In the case of marine adhesive proteins, however, total synthesis is not terribly practical. Some of the adhesive proteins have molecular weights upward of 100,000 mass units, which is currently inaccessible to fully synthetic proteins. Even when small proteins are synthesized, only very small quantities can be obtained, and at great cost.

Synthetic polymers, by contrast, can be made on enormous scales for very little cost. Think about how bulk polymers (i.e., "plastics") are all around you. The pen in your hand right now, the cover of the notebook in front of you, and the chair underneath you are all likely made from polymers. In terms of marine bioadhesives, we want to have the impressive material properties that these creatures design into their glues. We also want access to these materials in the same ways we have with bulk polymers or plastics. So, an intriguing approach is to take some of the chemistry from, say, a mussel adhesive protein and to incorporate it into a bulk polymer. This research direction is being explored by various labs, including our own. Let's walk through the thought process we are taking here.

Start with a mussel adhesive protein. We have a large molecular weight, maybe up to 100,000 mass units. We have the 20 standard amino acids plus DOPA, for 21 different monomeric units. That's pretty complex. In order to get a good adhesive, maybe all we really need is the long polymer backbone and, every now and then, a reactive group to mimic DOPA. This reductionist concept is shown in Figure 13.10, in which we think about a complex DOPA protein and then simplify it to any old polymer backbone with DOPA-like groups to keep the cross-linking chemistry that we know to be important.

For a simple polymer, we'll look at polystyrene. This particular polymer is easy to make on a large scale, has phenyl rings, which are somewhat similar in size to DOPA rings, is cheap, and has really no

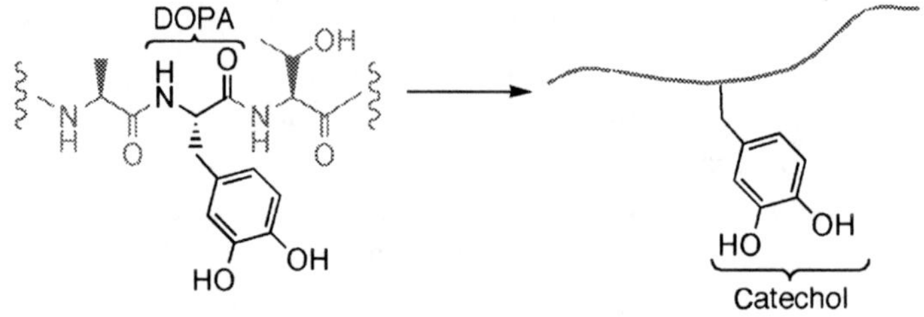

**Figure 13.10.** A DOPA-containing adhesive protein simplified to any polymer backbone with pendant catechol groups to mimic DOPA.

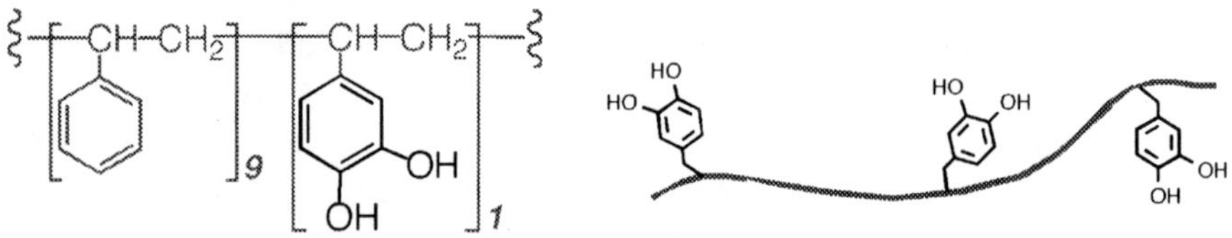

**Figure 13.11.** Biomimetic polymers based on polystyrene and modeled after mussel adhesive proteins. On the left is the chemical structure of a styrene and 3,4-dihydroxystyrene copolymer. On the right is a schematic showing that the catechol (or 3,4-dihydroxystyrene) groups are incorporated randomly throughout the polymer chain.

adhesive properties on its own. Then we will introduce catechol (i.e., 1,2-dihydroxybenzene; Figures 13.4 and 13.10) groups to the polymer to mimic DOPA in a protein. With mussel adhesive proteins, the amount of DOPA varies, but let's call it about 10% DOPA and about 90% other amino acid monomers. One of our synthetic mimics of mussel adhesive proteins is shown in Figure 13.11. Think about the polymer as a regular polystyrene with a reactive catechol group hanging off the chain every now and then, in random order. We may have nine styrene monomers in a row, then a monomer with the reactive catechol, then nine more styrenes, a catechol, and so on. We are making lots of different biomimetic polymers, but this polystyrene-based system serves to illustrate the thought process with a relatively simple example.

"OK, fine, but what does this fancy polymer do?" you may ask. Well, the first nice thing about this new polymer system is that we can make it on large scales—many grams or even kilograms at a time and with fairly little effort. More importantly, the polymer adheres well. Very well, in fact. Just like with the mussel adhesive straight from the animals, we can cross-link these polymers. Iron(III) works great as do some strong oxidants like sodium periodate ($NaIO_4$). The bulk adhesion is as strong as the cyanoacrylate "super" glues you may have used at home. That's really almost as strong as glues get. We even have the synthetic mimics adhering a little more strongly than the biological mussel glue after which it was designed. These new polymers can also stick strongly underwater.

So this example shows how we can explore materials made by nature, characterize the chemistry, and then use that knowledge to design new materials with interesting properties. So what's next? Let's try and develop some applications for these new materials. Which brings us to the next section ...

## SO WHAT'S IT GOOD FOR? APPLICATIONS OF MARINE BIOADHESIVES

OK, let's say that you have unlocked the secrets of the ocean. You know how your favorite sea critter sticks to rocks and you are also able to produce that material on a large scale. Now what?

Every time I go to see my dentist for a regular checkup, he asks "How's that glue of yours coming? Hurry it up, we're getting desperate here!" What he needs is a good dental cement. Something that will bond to your wet teeth strongly and stay there for years, without any troubles. Instead, what he has is a cement that does not work terribly well. The cement won't stick to wet teeth—you need to dry them rigorously before application. And even then, the current cements do not stick very well. They crack, crumble, promote cavities, and often do not last nearly as long as we would like. We definitely need something better: a really strong cement, something that can adhere to wet surfaces. Think about those barnacles we saw at the beach. They cement themselves to wet surfaces all the time, and quite strongly at that. With materials like barnacle cement or mussel adhesive, we might be able to bring the properties of marine biological materials to biomedical applications. The result could be new dental cements. Likewise, there is a need for good bone cements for surgical procedures filling voids, attaching metal implants to bone, and repairing various types of skeletal injuries.

Perhaps the most intriguing—and most difficult—application target for marine bioadhesives is that of a surgical glue. Each year, there are 230,000,000 major, invasive surgical procedures carried out worldwide. Nearly every one of these operations requires the use of sutures or staples. This typical surgical joinery is damaging in itself, can lead to infections, and delays healing, with the need for poking additional holes in the surrounding healthy tissue. Much better would be the ability to simply glue the internal tissues or skin back together. Surgical adhesives would also allow repair of fragile tissue, like that found in your intestines, which cannot be sutured. You might want to use a temporary glue that dissolves with time, after healing is complete. Or you might prefer the glue to stay there permanently. It depends on the specific application. Regardless, the world could really use an excellent surgical adhesive.

OK, let's think about the three major requirements you want in a surgical adhesive. First, you need to be able to set in a wet environment. Second, you want to make strong bonds. Third, your material cannot be toxic. Getting two of these three criteria in any given material is pretty easy. But combining all three is quite tough, so difficult, in fact, that no known material fulfills all three requirements. That's why we are looking to the seas. Think about, say, the mussel and barnacle adhesives. They stick to wet surfaces all the time. You saw that at the beach. You proved to yourself how strong these materials are when you pulled a few mussels off the rocks and couldn't get the barnacles to move. And after your time at the beach, maybe you stopped at a nearby seafood restaurant to have mussels for dinner. So these materials could be nontoxic. There you go: marine bioadhesives look like the perfect candidate materials from which we can develop surgical adhesives.

Now we need to get our hands on some of this stuff to test it out. In the previous section, we discussed various approaches to obtaining large quantities of marine bioadhesives for applications development. In our lab, we have decided to take the synthetic polymer approach (see Figures 13.10 and 13.11). Things seem to be going quite well. Take a look at Figure 13.12. Here we have used our new polymers to bond some biomedically relevant substrates. You can see an experiment in which we started with pig skin—an excellent model for human skin. We cut the skin into strips and, while still wet, applied our new polymeric adhesives. It all sticks together fairly well. Similarly, we are curious about dental and orthodontic applications for these materials. Figure 13.12 shows our polymers bonding a cow tooth to a strip of aluminum.

While we are on the topic of applications for marine biological materials, there are a few more that warrant mentioning. You're young, so you might have an old, rust bucket of a car. Wouldn't it be better if your car didn't rust? Heck, it would be great if we could stop metal from rusting in general. There may be an answer here, too, from the seas. Take a shiny piece of metal and put it into the water

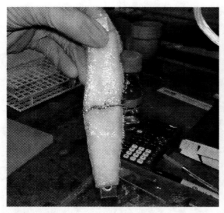

**Figure 13.12.** Using synthetic polymer mimics of mussel adhesives to bond together two strips of wet pig skin (left) and a cow tooth to aluminum (right). You can see the polymeric adhesive between the skin strips as well as between the tooth and metal.

at a nearby beach. Come back in a few weeks or months and you will find that the metal has rusted. Your panel is also covered with barnacles, mussels, and the like. Chip off the little critters. You will see that the metal is still shiny under there. So these marine bioadhesives could make for a new class of rustproof coatings.

Yet another application for these fascinating biological materials is in the area of tissue engineering. Let's say that you find yourself in need of a replacement organ—perhaps a liver, a kidney, or a heart. Where are you going to get one? It would be great if we could just grow you a replacement in the lab. Generally speaking, if you want to grow cells into a three-dimensional object like a heart, you will need a three-dimensional substrate onto, and throughout, which you can first deposit the cells. Then let the cells grow together and work in unison to become an organ. Recall that the glues of mussels and barnacles are made from cross-linked matrices of proteins. Such matrices may be the ideal environments within which we can grow cells. Some work in this area uses the actual protein adhesives from the animals. Other studies rely on synthetic polymer mimics. Either way, the future growth of replacement organs may have its origins in scaffolds from the seas.

Hopefully, you can now see that there is great potential for developing lots of exciting applications for technology derived from sea creatures. Be it a surgical adhesive, cements for bone or teeth, rustproof coatings, or ways to grow replacement organs. We're working in all of these areas. So come join us!

## STOP THE STICK: ANTIFOULING SURFACES

By any chance, do you have a boat? Nah, I didn't think so. But let's pretend that you do. Ahhh ... nice evening cruises into the sunset. Anyway, other than a leak, what is the one thing that you do *not* want on your boat? That would be a hull encrusted with barnacles, marine worm tubes, mussels, and seaweed. It's called marine fouling, and it will really slow you down.

Let's think about marine fouling from a global perspective. There are about 50,000 large commercial ships roaming the seas, including tankers, container vessels, and cargo ships. If these vessels were all crusted over with adhering marine organisms, the ships would consume an extra 70,000,000 tons of fuel each year. That's the equivalent to how much oil is used to power the United States for nearly a month! And this extra fuel use from fouled ship hulls could contribute around 200,000,000 tons of carbon dioxide into the air each year. So, if we can solve the fouling problem, we will have a significant impact on reducing climate change.

Take a look at the hulls of any large ship. Notice how they are very often painted red, down at the water line and below? That's an antifouling paint. It keeps barnacles, seaweed, and their friends from sticking to the ships. The reason the paint is red is that it's around 50% copper oxide by weight. Alkyl tin compounds such as tributyltin (TBT) used to be the main antifouling ingredient, but copper predominates now because it is a little less toxic. Both $Cu^+$ and $Cu^{2+}$ forms of the oxides are used in current paints. The painted hulls release copper ions ($Cu^+$ or $Cu^{2+}$) into the water. The larger ships can introduce nearly a kilogram of copper to the sea each day. When barnacles and the like are in their larval states, they cannot handle the high copper concentrations and die. As you might have guessed, dead barnacles do not go and attach themselves to ships. What all this means is that our current technology for preventing marine fouling is to simply kill what is in the water. Sure, it works. But this antifouling approach is a disaster from an environmental perspective. Go ahead and take a water sample from any harbor, bring it back to lab, and you will find elevated copper and tin levels there.

So how can we prevent marine fouling such that we will pollute the air less by using less ship fuel but without having to dump toxins into the water? Maybe you're thinking "Hey, why don't we coat the ship hulls with Teflon (polytetrafluoroethylene) because nothing sticks to that?" Well, look back at Figure 13.2, where we show a mussel sticking to Teflon. Barnacles can stick to Teflon, too. So the solution isn't that simple, unfortunately.

At the moment, there are not any really good antifouling paints, coatings, or surfaces that repel marine organisms without releasing toxins into the surrounding waters. Several labs are developing various approaches, many of which look promising. Some marine organisms prefer attaching to cationic (i.e., positively charged) surfaces, while others like anionic (i.e., negatively charged) surfaces. So, we can make surfaces that are simultaneously cationic and anionic ("zwitterionic") to confuse the organisms. Similarly, we can make surfaces that are both hydrophobic (i.e., water hating or oily) and hydrophilic (i.e., likes water or wettable) to confuse organisms that may like one surface property but not the other. Soft or rubbery surfaces seem to prevent a little adhesion relative to hard surfaces. Another approach is to zap the surfaces with electric pulses, thereby repelling the sticky guys. Yet another attempt at antifouling is simply making the surfaces taste bad to the critters. Coat the surface with an icky molecule such as capsaicin, the compound that makes peppers so hot, and you may be able to reduce the desire of foulers to be on your boat.

One of my favorite antifouling approaches, and one that our lab is pursuing, is to build upon the characterization studies described above. If we first can understand how these organisms stick, we then can devise surfaces that do not allow such adhesive interactions to take place. For example, look back at the iron and DOPA chemistry that we discovered (Figures 13.5–13.7). Those interactions between the metal and protein look to be key for mussels to make their glue. So what we are trying to do now is interrupt that process. If we can stop the formation of the glue, the animals will no longer be able to stick to your boat. Developing new antifouling coatings would be quite interesting, not to mention make a major, positive impact on the environment.

## YOUR FUTURE IN THE SEA

So there you have it: a wet and salty tour of chemistry at the beach! There's so much waiting to be discovered in the seas. I encourage you to come join us as we unravel the secrets of nature's materials!

Take it easy.

Jon

## FURTHER READING

Broad, W. J. Mapping the sea and its mysteries, *New York Times*, January 13, 2006, pp. D1 and D4.

Forbes, P. *The Gecko's Foot: Bio-inspiration: Engineering New Materials from Nature*, W. W. Norton and Company Publishers, New York, 2005.

Robbins, J. Second nature: More and more, innovative scientists are turning to the natural world for inspiration ... and design solutions. *Smithsonian* 2002, *33*, 78–84.

Smith, A. M.; Callow, J. A. (eds.) *Biological Adhesives*, Springer-Verlag Publishers, Berlin, 2006.

### Technical Articles

#### *Synthetic Polymer Mimics of Mussel Adhesive*

Westwood, G.; Horton, T. N.; Wilker, J. J. Simplified polymer mimics of cross-linking adhesive proteins. *Macromolecules* 2007, *40*, 3960–3964.

#### *Characterization of Mussel Adhesive*

Sever, M. J.; Weisser, J. T.; Monahan, J.; Srinivasan, S.; Wilker, J. J. Metal-mediated cross-linking in the generation of a marine mussel adhesive. *Angewandte Chemie International Edition* 2004, *43*, 447–450.

# 14

# The Advantage of Being Small: Nanotechnology

**Michael J. Sailor**

*University of California, San Diego*

Michael J. Sailor is a Professor of Chemistry and Biochemistry and the Leslie Orgel Scholar at the University of California, San Diego (UCSD). He holds affiliate appointments in the Departments of Bioengineering and Nanoengineering at UCSD. He received a BS degree in chemistry from Harvey Mudd College and a PhD degree in chemistry from Northwestern University. He joined the faculty at UCSD in 1990, after postdoctoral appointments at Stanford and Caltech. Professor Sailor is an expert in nanophase materials, with emphasis on silicon-based photonic systems. Current projects in his research lab at UCSD are directed at problems in nanoparticle-based diagnosis and treatment of disease, optical biosensors, detectors for toxins, pollutants, and biological warfare agents, energy harvesting and storage, and microfluidic systems.

Dear Angela,

I promised I'd send you some thoughts on nanomaterials chemistry. It took me a while to organize my thoughts, but I finally feel reasonably well prepared to give you a tour of the field, so here you go.

Nanotechnology was first proposed as a serious scientific discipline in the late 1950s. Caltech physicist Richard Feynman gave a lecture to a meeting of the American Physical Society titled "There's Plenty of Room at the Bottom." In that speech, Feynman invited people to enter what he considered a new field of science—the science of the very small. What started him thinking along those lines? Well, at that time, biologists were beginning to reveal the intricate machinery of microscopic organisms such as bacteria. The electron microscope and other recently developed tools were providing views into the complex mechanical structures and processes these creatures use to sense their environment, to move around, and to manufacture chemical fuels. At the same time, people were developing rudimentary methods to build very small devices—primarily electronic circuits for the microelectronics industry. Feynman reasoned that it was only a matter of time before we would design elaborate tools rivaling those nature uses to build its microscopic machines.

## THE ADVANTAGE OF BEING SMALL

One of my favorite images from that era comes from a book written by Dr. Seuss: *The Cat in the Hat Comes Back!* Published a year before Feynman's famous lecture, *The Cat in the Hat Comes Back!* involves a bunch of miniature cats produced from inside the hat of the Cat in the Hat. Figure 14.1 is an illustration from that book. The Cat in the Hat removes his hat to reveal a cat half his size standing on his head. Labeled "Cat A," that half-size cat in turn reveals "Cat B," half his own size under his hat, and so on, through the letters of the alphabet. So how small is that last cat, "Cat Z?"

Do you know that Dr. Seuss' real name was Theodore Geisel and that our university library is named for him? Just outside the library, there is a bronze statue of Geisel, with the Cat in the Hat standing just behind him looking over his shoulder. I measured that cat to be

Then Little Cat G
Took the hat off his head.
"I have Little Cat H
Here to help us," he said.

"Little Cats H, I, J,
K, L and M.
But our work is so hard
We must have more than them.
We need Little Cat N.
We need O. We need P.
We need Little Cats Q, R, S, T,
U and V."

**Figure 14.1.** *Illustration from Doctor Seuss,* The Cat in the Hat Comes Back! *Source: Beginner Books, distributed by Random House, New York, 1958.*

5 ft, 8 in. (1.7 m) tall. So let's assume that the statue is the height of the real Cat in the Hat. If you assume that each successive cat is half as tall as the previous one, you can calculate that Cat Z will be only $1.7\,m \times (1/2)^{26} = 2.5 \times 10^{-8}\,m$, or 25 nm tall. That's 2000 times smaller than the thickness of one of your hairs.

You might imagine that Seuss's smaller cats are not so special— their smaller arms and legs would make them slower and weaker than the larger cats. But as the story unfolds, you find that those small cats have capabilities the larger cats lack. They are brought in to help the children in the story clean up a pink stain covering the snow in their backyard. The small cats can get into places that the larger ones can't, but more importantly, they possess a material that doesn't exist in the "large" world:

Now don't ask me what Voom is.

I never will know.

But, boy! Let me tell you

It DOES clean up snow!

Resting inside the hat of Cat Z is a compound Seuss calls "Voom." This material creates a whirlwind that cleans up the stain in an instant. The fact that that last nanocat possesses something that performs new and different functions captures another essential feature of nanotechnology—when things get small, sometimes their fundamental properties change.

You can only cut something up so small before it loses its identity. Take a crystal of quartz, for example. Quartz is a mineral similar to glass, with the formula $SiO_2$. A pebble has the same color and hardness of the rock it came from, and if you grind the pebble into sand, the grains will be very similar to the pebble from which they came. But if you keep going, eventually, you separate the atomic constituents—the silicon from the oxygen in the case of quartz—and you no longer have the same thing you started with. The ancient Greeks were the first to understand this—they invented the word *atoma*, meaning "indivisible units" to describe the smallest piece of an object that you could possibly make. The basic concept of indivisibility eventually became a central concept of modern atomic theory, and our word "atom" descends from the Greek atoma. The Greeks could only do the thought experiment, but modern chemists break things up into molecules and atoms all the time. Modern physicists might say that the quark is the smallest indivisible unit obtainable, although the properties of a rock are long gone by the time you chop it up into quarks. So presumably, Cat Z was the atoma of cats—he was the smallest cat that could be made, because when he removed his hat, it held something utterly different from a cat. Although you can argue that an atom is the smallest chunk into which you can cut an element before it loses its properties, the properties of most things change when you get to the size of a molecule.

Let's look at an aspirin molecule, for example. It is an assembly of nine carbon atoms, four oxygen atoms, and eight hydrogen atoms (formula $C_9H_8O_4$), and it is 20 times larger than any one of the individual carbon atoms. The arrangement of these atoms in the bonds that make up the molecule determines its properties. If one of the bonds is broken—say, for example, the carbon bond holding the methyl ($CH_3$) group to the molecule—the material loses its painkilling ability. So for aspirin, the $C_9H_8O_4$ molecule is as small as you can go; it is the atoma of aspirin. These molecular units we chemists work with are pretty small—a molecule such as aspirin is only about 1 nm wide.

Aspirin

In your sophomore organic chemistry class this year, you will learn how to make compounds like aspirin, one bond at a time. The tools of chemistry can be very precise. However, the chemist's ability to make molecules such as aspirin does not constitute nanotechnology. You see, nanotechnology is not so precise as chemistry, and the "atoma" that determine the properties of nanomaterials are a bit larger than a molecule.

## SO WHAT IS NANOTECHNOLOGY?

The word nanotechnology comes from another Greek word, *nano*, or "dwarf." Like micro-, milli-, and mega-, nano- is a prefix we use to indicate a number multiplier. For example, *mega* is taken to represent the number 1 million, so a megaton is a million tons. Nano is one billionth, or $10^{-9}$. In the word *nanotechnology*, the prefix refers to the nanometer, a unit of length. When talking about nanotechnology, we are talking about things that are on the order of a few to maybe a hundred nanometers in size. That's at least 500 times smaller than the width of one of your hairs. Definitely, there's a certain size regime in which we nanotechnologists work, but it is not so much the specific size that's important here—it is the unique properties that derive from that size. GORE-TEX provides a great example of what I am talking about.

## STAYING DRY WITH NANOPORES

You should be familiar with GORE-TEX. It's a fabric that's used in outdoor clothing—rain jackets and boots and other garments designed to keep out the weather. The material itself is made up of a fibrous network containing very small holes, as shown in Figure 14.2. These

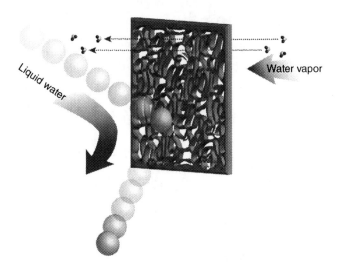

**Figure 14.2.** *Schematic view of a GORE-TEX membrane. The pores are too small to let liquid raindrops in, but they are large enough to allow individual water vapor molecules from perspiration out.*

pores in the fabric are about 100 nm wide. The amazing ability of a GORE-TEX membrane is that it repels liquid water, but it lets gas-phase water pass through. Thus, it keeps raindrops from leaking in while allowing your evaporated sweat to get out.

The magic of GORE-TEX is a nanoscale phenomenon: liquid water only forms droplets larger than ~100 nm. This size, known as the critical Kelvin radius, is determined by the surface tension of the liquid. The curvature at the surface of droplets smaller than the Kelvin radius is too great, forcing the droplets to either evaporate or to coalesce into larger drops. These same surface tension forces don't allow droplets of any size to squeeze through the 100-nm-diameter pores of the GORE-TEX fabric. However, individual molecules of water are less than 1 nm wide, and so water in its vapor form easily passes through the pores (see Figure 14.2). Therefore, when it rains on your GORE-TEX jacket, the water drops just sheet off. However, the water *vapor* that your body generates is able to pass through the jacket, keeping you dry. GORE-TEX truly is a nanotechnological material because it is the nanoscale size of the pores that provides the unique water-repellant yet breathable property—if the pores were much larger, liquid water would leak through, and if they were much smaller, your sweat would not evaporate and it would feel like you were wearing a plastic bag. That's really what I'm trying to get at when I tell you that nanotechnology provides us with new types of functions that depend on size.

GORE-TEX was invented in 1985. In my mind, it represents the first large-scale commercial application of nanotechnology. So why did it take so long for other nanotechnologies to emerge, and why does everyone seem to make such a fuss about it now as opposed to 25 or 30 years ago? There is one answer to both those questions: now we have a much more complete set of tools to build and examine nanostructures.

The most important tools are those that work at the molecular level. I already mentioned to you that molecules are made by chemists, and chemistry is not necessarily nanotechnology. However, the tools of the synthetic chemist are important to the field of nanotechnology. In your Chem 6A class, we demonstrated one of the simplest chemical reactions—the precipitation reaction. You probably didn't realize it when you heard about it in my lecture, but that chemical transformation is historically one of the most important reactions in the field of nanoscience.

## USING CHEMISTRY TO ASSEMBLE NANOSTRUCTURES

We demonstrated precipitation reactions in the context of the solubility rules of chemistry. The example we used was the reaction of silver ion with chloride ion in water. You may recall you learned that the ionic solid silver nitrate dissolves in water, but if a solution of sodium chloride (common table salt) is added to a silver nitrate solution, a fluffy white solid compound, silver chloride, forms. We usually write the equation for that reaction as

$$Ag^+_{(aq)} + Cl^-_{(aq)} \rightarrow AgCl_{(s)},$$

where the subscripts (aq) and (s) refer to an aqueous solution and a solid, respectively. A similar precipitation reaction occurs with cadmium ion ($Cd^{2+}$) and selenide ($Se^{2-}$). The precipitate in this case is the dark black solid, cadmium selenide (CdSe). Back in the early 1980s, researchers Arnim Henglein and Louis Brus developed methods to stop the precipitation reaction just as the solid was forming, resulting in individual crystals only a few nanometers in diameter.

The embryonic nanocrystals formed in the early stages of the precipitation reaction are not black like the larger crystals; instead, they can take on any color in the visible spectrum. The color of the

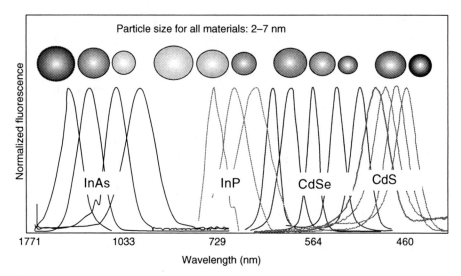

**Figure 14.3.** The fluorescence properties of nanosize semiconductors are size dependent. The plot shows the fluorescence spectra of a series of quantum dots of different sizes and compositions. Within a given composition series (InAs, InP, etc.), the color shifts to blue as the particles get smaller. Source: Bruchez, M.; Moronne, M.; Gin, P.; Weiss, S.; Alivisatos, A. P. Semiconductor nanocrystals as fluorescent biological labels. Science 1998, 281, 2013–2016.

crystals depends on the size of the nanocrystal. Furthermore, many types of these nanocrystals can fluoresce, and the fluorescence color is also size dependent. These size-dependent phenomena are referred to as quantum size effects, and they follow discrete quantum mechanical rules similar to those that govern electrons in atoms. For this reason, semiconductor nanocrystals such as CdSe are sometimes called "artificial atoms," or "quantum dots." The fluorescence spectra of a series of quantum dots, showing the size dependence of the emission maximum, are shown in Figure 14.3.

Quantum dots provide a great example of the enabling role that chemistry plays in the synthesis of nanomaterials. In that case, we harness the solubility rules to generate our nanomaterials. A simplified precipitation reaction for the synthesis of CdSe is shown in Figure 14.4. This type of nanoscale synthesis is often referred to as "bottom-up" because it starts with molecular or atomic building blocks. The other main approach to synthesize nanomaterials is called "top-down," in which we carve out a nanostructure from something larger. An example of this is found in the electrochemical synthesis of porous silicon nanostructures from crystalline silicon wafers.

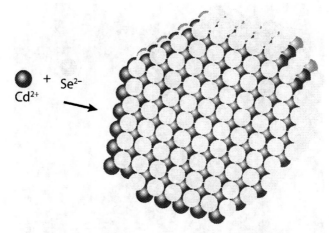

**Figure 14.4.** Semiconductor quantum dot nanocrystal synthesized by condensation of cadmium and selenium ions. This is an example of a bottom-up synthetic approach to construct nanostructures. In a bottom-up approach, the precursor atoms or molecules are combined to create a larger, hierarchical structure. In this case, cadmium ions react with selenium ions to form a cluster of the inorganic semiconductor cadmium selenide. Containing just a few hundred atoms, quantum dots are typically less than 7nm in diameter.

## USING ELECTROCHEMISTRY TO CARVE OUT NANOSTRUCTURES

Electrochemistry is a process in which we use electrons or electric current to either remove or deposit something onto a surface. You may be familiar with the metallic coatings placed on silverware or jewelry. Gold is an expensive metal, so the "14-karat gold" earrings you buy at those cheap stores at the mall are actually made mostly of steel that has a very thin layer of gold coated on the surface. The gold is deposited using an electrochemical process known as electroplating, and it can be very thin—on the order of just a few nanometers. With some materials, you can reverse the electroplating process to remove atoms from a surface. This process is called electrocorrosion. Corrosion is what causes your car to rust—it converts the iron to iron oxide. That process is spontaneous, but in some cases, the corrosion reaction can be driven with an electric current, with very precise results. For example, electrocorrosion of silicon can be used to generate very well-defined porous nanostructures. Figure 14.5 shows a cross-sectional view of a nanoporous, layered structure that was carved from a solid crystal of silicon. The silicon-based

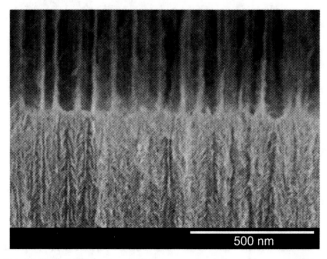

**Figure 14.5.** *Porous nanostructure carved from silicon. This is a side view of a silicon wafer that was "drilled" with nanoscale pores using electrocorrosion. The thickness of each layer and the size of the pores within a layer are controlled by the electric current applied during the electrochemical process. The method allows the "machining" of nanoscale structures with high precision. Image courtesy of Melanie L. Oakes, Hitachi Chemical Research Company.*

nanostructures can yield very intense and pure colors. The colors derive from a nano-optical phenomenon that is borrowed from a trick nature uses to generate the beautiful iridescent colors you see in peacock feathers, abalone shells, and the exoskeleton of many insects. What are these natural colors? Where do they come from? And how do we mimic them in artificial silicon nanostructures?

## COLORED NANOSTRUCTURES FROM NATURE

It turns out that nature has two ways of making color—one of them is to build a molecule, often referred to as a pigment. Pigments are made in nature's chemical factory, one bond at a time. The other method is to build a structure out of an otherwise colorless material. Rainbows have color due to the refraction of light in small, colorless water droplets. The spherical shape of a water droplet, and the angle between it and the sun, determines the colors you see. The rainbow of colors you see when you look at the surface of a DVD or a CD also comes from a structure—in this case, the fine, parallel grooves in the clear plastic disk. We call this structural color.

**Figure 14.6.** A polished abalone (paua) shell from New Zealand (left) and a crystal of the mineral calcite (right). Both materials are constructed from calcium carbonate, but in the abalone shell, the material takes the form of many layers, each a few hundred nanometers thick. These otherwise clear layers diffract light to impart the vivid colors seen in the shell.

The abalone shell and the plumage of a peacock are two organic examples of structural color. These organisms build color into their outer garments using physical structures rather than pigments. In the case of an abalone shell, the structural color derives from a sandwich of many layers of calcium carbonate. Calcium carbonate is a colorless material, commonly appearing in nature as the mineral calcite, shown in Figure 14.6. Whereas the calcite in that image is a clear, bricklike crystal, calcium carbonate in the abalone shell grows in thin layers, like the growth rings of a tree. Like our synthesis of the CdSe quantum dots I described earlier, the abalone uses a controlled precipitation reaction to form these layers; in this case, the precipitation reaction is of calcium ion and carbonate (Equation 14.1):

$$Ca^{2+} + CO_3^{2-} \rightarrow CaCO_3(s). \qquad (14.1)$$

Figure 14.7 shows an electron microscope image of these layers from a shell that was cut in half to reveal the cross section. Light passing through these layers is diffracted into specific colors, or wavelengths, resulting in the intense and beautiful green, blue, and violet colors you observe when you look at a polished shell. Fragments of abalone shells are often used in jewelry because of these striking colors. You have probably seen many other examples of structural colors without realizing what you were looking at. The iridescent

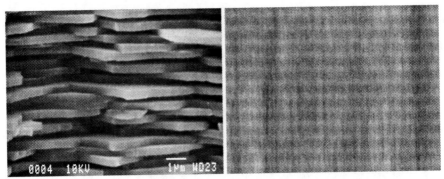

**Figure 14.7.** *Electron microscope images comparing the layered nanostructures of an abalone shell (left) and a porous silicon nanostructure (right). The even layer-to-layer spacing of a few hundred nanometers diffracts specific colors of light, producing beautiful red, green, blue, or violet hues. Left image from Morse, D. E.; Cariolou, M. A.; Stucky, G. D.; Zaremba, C. M.; Hansma, P. K. Genetic coding in biomineralization of microlaminate composites. Materials Research Society Symposium Proceedings 1993, 292, 59–67. Right image courtesy of Melanie L. Oakes, Hitachi Chemical Research Company.*

appearance of many birds and insects arises from the structural color phenomenon.

## SILICON-BASED PHOTONIC CRYSTALS

The structurally colored abalone shell is an example of what the optical physics community calls a "photonic crystal" because the repeating, uniformly spaced layers act as a crystal, diffracting photons of light. We can build the same kind of color into artificial structures, using the electrochemical silicon machining method I mentioned earlier. How we make these layered structures in silicon is an example of a top-down approach used to make nanostructures. Whereas the abalone shell is synthesized bottom-up by the precipitation of layer upon layer of calcium carbonate, the layered structure in silicon is formed by drilling holes in a solid crystal of silicon.

When we immerse a silicon wafer in a solution of hydrofluoric acid and apply electric current, the silicon wafer dissolves. However, it doesn't dissolve uniformly. The electric corrosion process drills millions of tiny nanometer-scale pores into the wafer. The diameter of the holes is determined by the electric current, so their size can be varied during the etch. If we turn the current up and down, the propagating holes widen and narrow accordingly. The two-layer porous structure in Figure 14.5 was prepared this way. The pores were drilled from the top to the bottom of the image, starting with a large current and reducing it halfway through the etch to generate the sharp decrease in pore size. If the current is repeatedly cycled, the multi-layer structure of Figure 14.7 results. These alternating layers of large and small pores diffract light similar to the layers in the abalone shell, yielding a manufactured photonic crystal with colors as vivid as the natural counterpart. The cross-sectional electron microscope images of Figure 14.7 are both layered photonic crystals, though one is carved out of silicon and the other is constructed from a successive deposition of calcium carbonate layers.

Although the colors that are generated from these layered silicon materials are similar to the colors of the abalone shell, there are some interesting things you can do with the silicon-based nanostructures that can't be done with the natural ones. Unlike the abalone shell, the layers in the silicon nanostructures are porous, and molecules can enter these hollow materials. When that occurs, the color of the nanostructure changes. For example, a green porous silicon photonic crystal will turn red when it is immersed in a liquid. This color change has been harnessed to make sensors for chemical compounds. What's happening here is the silicon structure is filling up with molecules that have a refractive index different from air, and the color of a photonic crystal is determined by the refractive index of its components. Various highly sensitive chemical or biological sensors have been built based on this simple type of color change.

I've given you an example of how we can build nanostructures in silicon that have unique colors and how we can harness those colors to make a simple sensor. I'll finish with an example of how the nanoscale phenomenon that keeps liquid water from penetrating GORE-TEX can be applied to the pores in the silicon nanosensor. Figure 14.8 shows two porous silicon photonic crystal samples that were prepared with the same nanostructure, but they have different chemical treatments applied to their surface. To each sample in the image we applied a drop of water. You can see that the water drop beads up on the sample on the left. This is an example of the surface tension-driven rejection of water I described earlier for GORE-TEX. The water drop is more spread out on the sample at the right. That chip contains a hydrophilic surface coating, a thin layer of $SiO_2$, which allows water to wick into the pores. The left-hand sample contains a more hydrophobic surface chemistry (dodecane) that completely excludes water from the nanopores. The lower image is a side view that clearly shows the affinity of each surface for

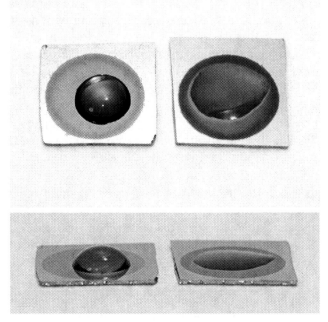

**Figure 14.8.** *Two views of a pair of porous silicon photonic crystals, each with a drop of water placed on the surface. The sample on the left contains a hydrophobic surface chemistry (dodecane) that excludes water, while the sample on the right contains a hydrophilic surface (SiO₂). Water spreads out on the surface and wicks into the hydrophilic chemistry, shifting the color of the photonic crystal to red. The lower image is a side view showing the affinity of each surface for water.*

water; the more spherical shape of the water drop on the left demonstrates the "beading up" that occurs on hydrophobic surfaces. When the water drop wicks into the hydrophilic $SiO_2$ layer, the color of that photonic crystal shifts to the red as the high refractive index of the invading water molecules pushes the photonic band to lower energy.

The different behavior of the two chips shown in Figure 14.8 demonstrates the important role that chemistry plays in nanotechnology. The chemistry in the sample on the left consists of a long-chain hydrocarbon—essentially an oil—that repels water and forces it to bead up. The two chips were prepared with the same nanopore morphology; it is the surface chemistry that serves as a gatekeeper to control the admission of the chemicals we want to sense. The surface chemistry assists the nanopores in deciding when to reject and when to admit water. It turns out that GORE-TEX requires a similar hydrophobic chemical assistant; in that case, the hydrophobic molecules are incorporated into the molecular structure of the polymer used to make the fabric. Without that chemistry, GORE-TEX would wick up water just like a sponge.

## IRON OXIDE NANOPARTICLES TO DETECT TUMORS

Nanotechnology applied to the medical sciences is being called "nanomedicine," whose goal is to provide better methods to administer drugs or to detect diseases. For example, researchers in the field of cancer detection have been developing nanoparticles based on iron oxide, which can be injected into a patient to allow doctors to find cancerous tumors more effectively. Tumors in their early stages of growth can be difficult to detect by any method. However, it is at this early stage when the disease, if caught, is more likely to be cured. One of the most powerful tools used to find tumors in humans is magnetic resonance imaging (MRI). A common tool used by doctors to look inside the body, MRI is particularly sensitive to magnetic materials, and magnetic iron oxide nanoparticles such as magnetite ($Fe_3O_4$) are useful in this regard. One long-standing goal has been to prepare iron oxide nanoparticles with the right surface chemistry to allow them to swim through the bloodstream and then to find and stick to tumor cells or other tissues that associate with tumor cells. This enhances the contrast of the images, allowing doctors to see smaller tumors by MRI, hopefully enabling the diagnosis of cancer at earlier stages.

The synthesis of iron oxide nanoparticles follows a bottom-up precipitation reaction similar to the reactions used to make CdSe nanoparticles and $CaCO_3$ layers in seashells. In this case, it is the reaction of $Fe^{2+}$ and $Fe^{3+}$ ions with a basic aqueous solution:

$$Fe^{2+}_{(aq)} + 2Fe^{3+}_{(aq)} + 8NH_{3(aq)} + 4H_2O \rightarrow Fe_3O_{4(s)} + 8NH^+_{4(aq)}. \quad (14.2)$$

Iron oxide nanoparticles can take on various shapes and properties, depending on the intended medical application. Figure 14.9 shows an image of "nanoworms" prepared by stringing several iron oxide nanoparticles together in a chain using a polymer. Nanoworms have shown significant improvements in tumor targeting tumors relative to spherical iron oxide nanoparticles. The reason these worms work so well is due to a combination of their surface chemistry and their nanoscopic shape.

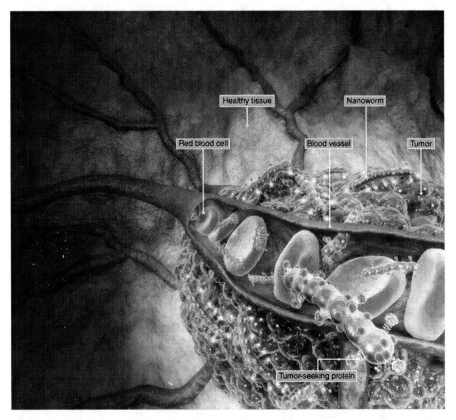

**Figure 14.9.** *Artist's depiction of iron oxide nanoworms finding a tumor in the body. The worms are made of magnetic iron oxide coated with a biopolymer. The wormlike structure and a specialized coating allow these nanodevices to find and attach to tumors, improving our ability to find the tumor in an MRI scan of the patient. Image from* Popular Science, *November 2008.*

## USING CHEMISTRY TO IMPROVE NANOPARTICLE PERFORMANCE: ANCHORING NANOPARTICLES TO TUMORS

Most nanoparticles are recognized by the body's protective mechanisms as foreign objects. They are captured and removed from the bloodstream within minutes. A common trick is to coat the nanostructure with a biopolymer that allows the nanostructure to evade these natural elimination processes. In the case of iron oxide nanoworms, the biopolymer coating allows the materials to circulate in the body of a mouse for hours.

To the biopolymer coatings we can add specialized molecules that home to tumors. A wide range of such tumor-targeting molecules have been discovered and synthesized by researchers like Professor Erkki Ruoslahti, at the Burnham Institute at the University of California, Santa Barbara. When attached to a nanostructure, these molecules act like glue to stick the nanoparticle to a tumor or they can perform more complex tasks, like transporting the nanoparticle across the cellular membrane where it can more effectively kill the tumor.

## THE GOOD, THE BAD, AND THE UGLY OF NANOTECHNOLOGY

Figure 14.10 shows covers from two books that highlight the challenges we face in nanotechnology. On the left is the cover of the

**Figure 14.10.** Contrasting views of the future of nanotechnology. Covers from the novels Fantastic Voyage (© 1966, by Isaac Asimov) and Prey (© 2002, by Michael Crichton).

novel *Fantastic Voyage*, written by Isaac Asimov back in the 1960s. The other image is from *Prey*, written by Michael Crichton almost 40 years later. The scientists in *Fantastic Voyage* develop a technology to shrink a team of people in a submarine to the size of a blood cell. When injected into the body of a comatose patient, the submarine made its way through the circulation to find and remove a dangerous blood clot. That really captures the essence of what we're trying to do with medical nanomaterials in the nanoworm example I just discussed. Though we aren't trying to shrink *people* down to that size, we are building the submarines, at least in the form of small devices that can circulate through the body, and we're trying to give those structures some capabilities to perform complex tasks to cure patients. So in a broad sense, that book cover represents nanotechnology being applied for the benefit of mankind.

The other cover, from the book *Prey*, is "nanotechnology gone bad." In that story, malevolent nanodevices try to take over the world. Although both of these stories are science fiction, they bring up the very real point that any technology can have good or bad consequences. With a new technology, these consequences are often not appreciated until after the cat is out of the bag, so to speak.

I know you thought you were done getting quizzes from me when you finished my Chem 6A class, but I have one more for you. Take a look at the list below and identify the formula of a deadly chemical.

Quiz: Which is the deadly chemical?

(a) $H_2O$
(b) $NaCl$
(c) $Fe_2O_3$
(d) $SiO_2$
(e) $Na_2O \cdot Fe_2O_3 \cdot 3FeO \cdot 8SiO_2 \cdot H_2O$

Well, the answer to this quiz you can probably guess by a process of elimination. Answer (a) is of course water; answer (b) is sodium chloride, or table salt; answer (c) is one form of iron oxide, or rust; and answer (d) is silicon dioxide, the formula for sand and glass. So what is left is answer (e), which is the formula for the mineral crocidolite, a form of asbestos that causes the deadly lung cancer mesothelioma.

Answer (e) has a complicated formula, but if you look at it closely, you'll see that it seems to be made of some pretty harmless things:

**Figure 14.11.** *Nanofibers of the natural mineral crocidolite—one of the most toxic asbestos minerals.* Source: *Longo, W. E.; Rigler, M. W.; Slade, J.* Cancer Research *1995, 55, 2232–2235.*

iron oxide, silicon dioxide, oxygen, and hydrogen. Its not as if this compound is made of plutonium or some other toxic elements—in fact, the elements themselves are quite harmless. It's the form, the nanostructure of this material, that makes it so dangerous. An electron microscope picture of crocidolite asbestos fibers is shown in Figure 14.11. You'll notice that it consists of a bunch of very fine fibers only a hundred nanometers wide. These fibers are so stable that when they get into the lungs, they can stay there for more than 20 years. Over time, the small fibers cause irritation and damage to the lung tissues, which can lead to cancer and other diseases.

That's why asbestos is so harmful; it's not the elements that comprise the material, it's the form of the material—the fine, stable fibers.

Our experience with asbestos underscores perhaps the most important aspect of nanotechnology development, that the material needs to be harmless or it needs to dissolve away into something that's harmless if it is going to be used in the body. Furthermore, all industrial materials eventually find their way into the environment, and so similar precautions need to be considered for the nonmedical uses of nanomaterials.

Degradability is one of the hallmarks of a lot of the work that many of us do in the field of nanotechnology. Perhaps it isn't as important if only a very small amount of the nanomaterial is to be locked inside a computer chip, but if the material is going to be spread on the roof in the form of solar panels, or if it is to be incorporated into paints and ceiling tiles as was done with asbestos, we really have to be careful in handling and disposing of the material. Or we have to design into our nanomaterials the ability to self-destruct after they have performed their duties. So for us working in nanotechnology, the chemical reactions of the material are as important as the chemical constituents.

In addition to *Fantastic Voyage*, science fiction writer Isaac Asimov wrote a series of short stories about futuristic robots. In the *I, Robot* series, he introduced what he called the Three Laws of Robotics to define the rules that all robots were required to follow. These rules, the first of which was "A robot may not injure a human being or, through inaction, allow a human being to come to harm," were incorporated into the fundamental programming of every robot. Many people are now struggling with the question of how to regulate nanotechnology to minimize the unintended, harmful consequences of this very new field.

## LAWS OF NANOROBOTICS

I think a set of rules needs to be incorporated into the chemical programming of every nanorobot.

(1)    The device must not self-replicate in its host environment.
(2)    The device must degrade in its host environment.
(3)    The degradation products must be harmless to the host environment at the concentrations at which they are produced.

Keep in mind that "harmless" is a very difficult term to define. If you ask the question "is carbon dioxide harmless," most people would probably say yes. You exhale carbon dioxide with every breath, and it is needed by plants to live and grow. But animals or humans are asphyxiated if placed in a high concentration of carbon dioxide, and now we are finding that very small increases in carbon dioxide concentrations in the atmosphere can change our climate. It would be hard to program Asimov's first law of robotics into a nanostructure because the device cannot be sophisticated enough to know what is harmful and what is harmless to its host (even we humans have a hard time with that concept). Clearly, the concentration of nanodevices is important, which is why I prefer that we build the devices ourselves rather than allow them to replicate themselves without intervention, like a virus. You also need to know what the "host environment" is (a person, a lake, a building, etc.) and how it might be harmed by the technology. Can you think of some more rules to apply to nanorobots?

I'd like to finish where I began with the picture from *The Cat in the Hat Comes Back!* in Figure 14.1. The reason this image is important in describing nanotechnology is that it captures the essence of what we're trying to do. We calculated that smallest cat in the smallest hat to be only 25 nm tall. But it isn't so important to make it small. We have to give that little cat eyes, a nose, a mouth, and hands so that it can do things bigger cats can't. The goal of nanotechnology is to develop those tools—to construct operational devices a million times smaller than a cat.

Oh, well, I see I got rather carried away. That was by far the longest letter I have ever written (even allowing for the fact that I belong to a generation that wrote long letters once in a while). Feel free to share this with any friends who might be interested and do stop by my office for a chat, if I can be of further assistance.

Best,

Mike Sailor

## ACKNOWLEDGMENTS

The author thanks Gitfon Cheung, Daniel E. Morse, Luo Gu, Sarah M. Cheng, Manuel M. Orosco, Melanie L. Oakes, and Gordon M. Miskelly for their contributions and comments.

## FURTHER READING

Color images of the figures shown in this chapter are available at http://sailorgroup.ucsd.edu/research/images.html.

Austin, R. H.; Lim, S.-F. The Sackler colloquium on promises and perils in nanotechnology for medicine. *Proceedings of the National Academy of Sciences of the United States of America* 2008, *105*, 17217–17221.

Feynman, R. P. There's plenty of room at the bottom: An invitation to enter a new field of physics. A transcript of the classic talk that Richard Feynman gave on December 29, 1959, at the Annual Meeting of the American Physical Society. 1959. Available at http://www.zyvex.com/nanotech/feynman.html.

Goodsell, D. S. *Bionanotechnology: Lessons from Nature*, Wiley-Liss, Hoboken, NJ, 2004.

Park, J.-H.; Maltzahn, G. A. V.; Zhang, L.; Schwartz, M. P.; Bhatia, S. N.; Ruoslahti, E.; Sailor, M. J. Magnetic iron oxide nanoworms for tumor targeting and imaging. *Advanced Materials* 2008, *20*, 1630–1635.

Ratner, M. A.; Ratner, D. *Nanotechnology: A Gentle Introduction to the Next Big Idea*, Prentice Hall PTR, Upper Saddle River, NJ, 2003.

Rubahn, H.-G. *Basics of Nanotechnology*, 3rd Edition, Wiley-VCH Verlag GmbH, Weinheim, 2008.

Sailor, M. J.; Link, J. R. Smart dust: Nanostructured devices in a grain of sand. *Chemical Communications* 2005, 1375–1383.

Steigerwald, M. L.; Brus, L. E. Semiconductor crystallites—A class of large molecules. *Accounts of Chemical Research* 1990, *23*, 183–188.

Wilson, M. *Nanotechnology: Basic Science and Emerging Technologies*, Chapman & Hall/CRC, Boca Raton, FL, 2002.

*Part IV*

# *Chemistry and Energy*

# 15

## Happy Campers: Chemists' Solutions to Energy Problems

**Penelope J. Brothers**

*The University of Auckland*

Penny Brothers was born and grew up in Auckland, New Zealand, and completed BSc and MSc(Hons.) degrees in chemistry at the University of Auckland. In 1979, she was awarded a Fulbright Fellowship and set off for Stanford University to begin a PhD in chemistry under the supervision of Professor Jim Collman. Her PhD thesis, and much of her subsequent research work, has centered around the chemistry of porphyrin complexes. In 1986, she returned to Auckland and spent 2 years working as a postdoctoral fellow with Professor Warren Roper in the Department of Chemistry, focusing on organometallic chemistry. In 1988, she took up her current academic position at the University of Auckland and was promoted to professor in 2009. She has been a visiting scientist at Los Alamos National Laboratory (2003, 2005, and 2007) and a visiting professor at the University of California at Davis (1993), the University of Heidelberg (2003), and the University of Burgundy (2004 and 2006).

*Letters to a Young Chemist*, First Edition. Edited by Abhik Ghosh.
© 2011 John Wiley & Sons, Inc. Published 2011 by John Wiley & Sons, Inc.

She was awarded a Fulbright Senior Scholar Award to visit Los Alamos National Laboratory in 2007. She is a member of the *Chemical Communications* Editorial Board. Her current research brings together her interests in porphyrin chemistry, the main group elements, and organometallic chemistry. She investigates how porphyrin and corrole ligands can be used to modify the chemistry of elements such as boron and bismuth.

Dear Angela,

I enjoyed meeting you when I visited the University of California, San Diego (UCSD) to give a seminar a few months ago. I think it's a great idea to invite a couple of undergraduate students to join the speaker for lunch. One of the things I enjoy the most about visiting other universities is the chance to meet students. I'm also impressed that you remember me—I guess women chemistry faculty are still, sadly, a bit of a rarity, and that fact, together with my Kiwi accent, may have helped. Hopefully, by the time you are in the middle of your future career, women chemistry faculty will be taken for granted.

You asked a bunch of interesting questions about sustainability in your e-mail message last month, and I promised to tell you a bit more about some of the work I do. Now that my summer vacation is over, I have time to sit down and write. Speaking of summer vacation, here in New Zealand, that means December and January, and I spent the December and New Year holidays camping at the beach with my family. You mentioned that you liked camping with your family when you were growing up, and so I thought that might be a good context to start talking about what interests me.

Right before I went on vacation, I attended a "Sustainability in Science and Engineering" conference here in Auckland. The theme of the meeting was how the work of scientists and engineers will help to create a sustainable future. So I had all those ideas in my mind as we set off for our camping trip. Camping is a pretty low-tech, sustainable kind of vacation (so long as you do it in a tent, not an enormous RV), but I got to wondering about how things might be different in a truly sustainable future and how chemists like me (and you too, if it turns out to be something that excites you) will help us get there.

We started out with several hours' drive to the beach, our car loaded down with four of us and all the gear. Car = gasoline = fossil fuels = carbon dioxide. Whew! That's a bad start. We pitched the tent (nylon), blew up the air mattresses (vinyl), and set out our coolers and camp furniture (aluminum and more plastics and synthetic fabrics). That's another load of materials sourced ultimately from fossil fuels, except for the aluminum, but it requires a huge amount of electricity for its production. We are pretty comfortably set up with a gas camp stove and gas lantern for cooking and lights—oops, more fossil fuels, more $CO_2$.

So what do we need for a more sustainable camping vacation? We need a better source of energy for fuel—for transport, cooking, and lighting, and for producing metals like aluminum—and better materials (plastics and fabrics) for the gear that we use. The materials need to come from a sustainable source and should either be recyclable or break down to benign products once their useful life is over (how many people do you know who have garages full of old, broken camping gear?). Maybe the materials will also be "smart" enough to do more than one job at once—imagine if the tent fabric could collect sunlight energy and convert it to electricity as well as shelter you from wind and rain. So those are big challenges already, and that's just for a low-tech camping trip! Imagine if you ask the same kind of questions about more complex enterprises—airplane travel, a shopping mall and everything in it, a manufacturing plant, an iron mine—and you can see that sustainability issues are huge, challenging, and vitally important.

## ENERGY: CHEMISTS HAVE SOLUTIONS

I've touched on two main themes for my utopian, sustainable camping vacation: energy and materials. I think I'll talk about materials another time and talk mostly about energy here. I firmly believe that the solutions to the world's energy problems will be delivered primarily by chemists (with help from all kinds of other scientists and engineers, but the fundamental breakthroughs will be in chemistry). I'll take a few minutes to defend that point of view, then tell you a bit about the kind of chemistry I do, and finally explain some of the ways it might be useful in addressing energy problems.

Of the truly sustainable energy sources—solar, wind, hydro, geothermal, and tidal—solar energy produced directly from sunlight currently offers the most promise. The other sources are not evenly distributed over the planet, and some require major infrastructure. Every year, the sun delivers, completely free, 10,000 times more energy than the global population currently consumes. Of course, in order to be usable, we need to convert the solar energy to a more accessible form, usually either thermal energy or electricity. Solar thermal energy uses the energy of sunlight to raise the temperature of a substrate material, usually a fluid. This can be water (heating swimming pools or domestic hot water) or, using more sophisticated designs, other fluids that can be heated to higher temperatures, and the heat energy is then converted to electricity.

Photovoltaics, on the other hand, convert sunlight directly to electricity by triggering a chemical change in the material absorbing the sunlight. This chemical change—how to induce it, how to collect the electrical current, how to understand which materials can be used, and how to make the process more efficient—are all fundamentally chemical questions and will require chemists to answer them. Chemists are the scientists who have the deepest understanding of the key first step, the chemical change that creates a free electron that is drawn off as the electrical current, and have the skills to design and make molecules and materials with exactly the right properties to optimize the efficiency of the process. Using the tools of the emerging field of green chemistry, chemists can also make sure that past mistakes are not repeated and that the new systems they design will utilize sustainable materials and processes.

The chemical changes that occur when sunlight is absorbed can be used to create electricity, which can be used directly as a power source. An example would be a solar-powered car that has collectors on the roof converting sunlight to electrical energy, which is stored in batteries, and the car itself is battery powered. This kind of technology is already possible, although it has low efficiency as only about 10% of the incoming solar energy can be captured and converted. On a larger scale, the electricity could be used for the aluminum refinery, which produces the metal used to make our tent poles and camp furniture. A more futuristic application would be a smart tent made from fabric coated with flexible material that was also a solar collector creating electricity to charge batteries, which could be used for light (and even sound—plug your iPod into your tent). But this is future technology and will require chemists to invent it. We'll look at how this might be done a little later on, but in the meantime, we'll investigate another use for the electricity produced from sunlight.

# SUNLIGHT + WATER = HYDROGEN?

Another way to use solar electricity is to drive an uphill chemical reaction to produce a new fuel. The classic example of this is water splitting. Water, better known to chemists as $H_2O$, has constituent oxygen and hydrogen atoms. As pure elements, they exist as hydrogen and oxygen gases, $H_2$ and $O_2$. Hydrogen is extremely flammable (remember those old pictures from early in the twentieth century of airships blowing up). When ignited, mixtures of hydrogen and oxygen react exothermically (giving off a lot of heat energy) and form water as the product. This is a "downhill" chemical reaction; in other words, it gives off energy. The reverse reaction, splitting water, $H_2O$, to form hydrogen and oxygen, $H_2$ and $O_2$, is an "uphill" chemical reaction, which means we have to add energy to make it happen. We already know that this can be done—using electricity as the source of energy together with a catalyst (Figure 15.1). The catalyst creates a pathway for the reaction to occur but doesn't itself get consumed in the reaction. However, currently, the best catalyst is platinum metal, which is expensive and relatively rare, so while it can be used on a small scale, it might not be practical—or affordable—to use on a really large scale.

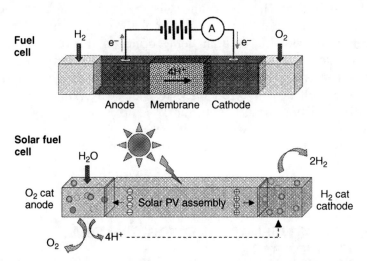

**Figure 15.1.** *A solar fuel cell uses light energy to drive the reaction producing hydrogen and oxygen from water. A fuel cell uses the reaction combining hydrogen and oxygen to form water to produce electrical energy. Reproduced from Lewis, N. S.; Nocera, D. G. Powering the planet: Chemical challenges in solar energy utilization. Proceedings of the National* Academy of Sciences of the United States of America *2006, 103, 15729–15735.*

## HYDROGEN: A FUEL FOR THE FUTURE

Why is this such an important process? Water is abundant on this planet, safe, nontoxic, and essential to life. Equally important, it is also renewable. If we have a free source of energy (sunlight) that can be used to drive the uphill reaction to form hydrogen and oxygen from water, then we have effectively stored the energy in the chemical bonds in $H_2$ and $O_2$. If we carefully control the reverse, downhill reaction of $H_2$ with $O_2$ so that we can use the energy that is released, we are essentially using hydrogen as a fuel. Oxygen, the other reactant, is readily available in the atmosphere. The only by-product is water—safe, nontoxic, and essential to life—no volatile hydrocarbons or fuel additives, no $CO_2$ being produced, no depletion of fossil fuels. It sounds like a dream—and it's a dream that many people already have. Use sunlight energy to create electricity; use the electricity to produce hydrogen from water; use the hydrogen as a fuel by recombining it with oxygen to release energy; and produce water as the only by-product.

Of course, if it was simple, it would have already been done. I mentioned the need for a cheaper, more readily available catalyst for the step required to produce hydrogen and oxygen for water. Very recently, there has been an exciting breakthrough by chemists at MIT who have discovered that cobalt oxide can do the job. There is also a lot of research being done on the reactor in which the controlled reaction of hydrogen and oxygen occurs, known as a fuel cell. The energy released by the reaction of hydrogen and oxygen in the fuel cell is converted back to electricity. This also requires a catalyst (again, usually platinum). There are already examples of hydrogen-powered vehicles in use, but it's not mainstream technology yet, and in addition, the hydrogen they use comes from fossil fuels.

## A FUEL TANK WHEN IT'S NOT FOR GASOLINE: HYDROGEN STORAGE

Although, as I have outlined above, technology for making hydrogen from water using electricity and fuel cells for the reaction of hydrogen with oxygen to produce water and electricity are already known, there are still some big challenges to tackle before we can all drive around in hydrogen-powered cars. First, think about how gasoline works. At room temperature and pressure, gasoline is a liquid that can be poured or pumped. You can take the cap off your car gas tank and the gasoline won't all escape. It's stored on a large scale (ranging from refineries to gas stations) and can be conveniently transferred into smaller units for storage near its point of use (the fuel tank in your car). How would this work for hydrogen? If we look into the properties of hydrogen a bit more, we see that the issues of hydrogen storage are very different than those of gasoline, and there are some big issues to resolve before we can conveniently pull into a "hydrogen station" for a fill-up.

Hydrogen is very different from gasoline. It is a lightweight (H is the lightest atom in the periodic table), explosive gas that needs to be stored and handled under pressure. To better compare the characteristics of gasoline and hydrogen, we need to know about energy density. This is the amount of energy stored in a given system or region of space per unit volume, or per unit mass. Hydrogen has a higher energy density per unit mass than does gasoline but a much lower energy density per unit volume (Figure 15.2). In other words,

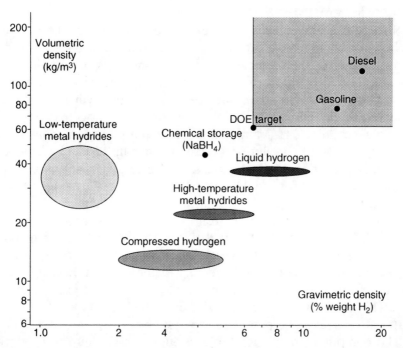

**Figure 15.2.** Density of hydrogen by volume and by weight in hydrogen-containing materials. Reproduced from van Noorden, R. Hydrogen storage targets out of reach. Chemistry World 2007, 4(10), http://www.rsc.org/chemistryworld/Issues/2007/October/HydrogenStorage TargetsOutOfReach.asp.

a kilogram of hydrogen contains more stored energy than a kilogram of gasoline, but we require a much higher volume to store a kilogram of hydrogen. This means hydrogen tanks need to be large, and pressure equipment (think stainless steel tanks) tends to be very heavy, which means that to carry hydrogen stored in gaseous form on a vehicle, some proportion of the hydrogen energy is used just to move the storage equipment around.

Are there alternatives? Remember, we have to think about not just the energy density of different forms of hydrogen, but also about other energy costs that might be involved. Every additional energy cost in the system reduces the amount of energy available for the end use. For example, compressed hydrogen requires energy to power the compressor. Although higher compression will result in higher energy density, more energy will be lost in the compression step. Liquid hydrogen is a possibility, but hydrogen must be cooled to −253°C (20 K) in order to liquefy it, requiring a lot of energy, and stored in expensive, well-insulated tanks. Even then, the energy density of liquid hydrogen is still four times lower than in gasoline: there is actually about 64% more hydrogen in a liter of gasoline (116 g hydrogen) than there is in a liter of pure liquid hydrogen (71 g hydrogen).

If there is a higher volumetric energy density in gasoline (a hydrocarbon containing both carbon and hydrogen) than there is in pure liquid hydrogen, then we can consider other kinds of chemical compounds as hydrogen storage materials. They need to contain lightweight elements (so that the percentage of hydrogen remains relatively high) and to be able to reversibly release hydrogen. For example, hexane ($C_6H_{14}$) and octane ($C_8H_{18}$) are hydrocarbons similar to components of gasoline and contain around 16% hydrogen by weight. However, the U.S. Department of Energy has set targets (percentage by weight) for the development of on-board hydrogen storage materials, which include the weight of all the containers and components of the storage system as well as the actual hydrogen-containing material. The targets are 6% by 2010 and 9% by 2015. As an example, if a hydrogen storage material itself contains 16% hydrogen, but the system required to store and use 5 kg of the compound itself weighs an additional 5 kg, then the overall hydrogen content is calculated as 8%. This rules out using heavy chemical elements in the storage material or heavy equipment in the system.

The compound must be stable enough for medium-term storage (won't release hydrogen before it is needed) but must be reactive enough that it can release hydrogen under mild conditions when it is required. It won't be useful if the hydrogen release step requires a lot of energy to initiate. There is also the spent fuel to consider. The spent fuel is not an issue for gasoline—gasoline itself is the fuel and when it's all burnt, it's all gone (converted to $CO_2$ and water and released to the atmosphere). When hydrogen is the fuel and a chemical compound is the storage medium, once the hydrogen is released, the new chemical compound resulting from loss of hydrogen (the spent fuel) remains. It must be able to be regenerated (add the hydrogen back) without a high overall energy cost. All the boring thermodynamics you learned in P-chem starts to become more interesting when there are real applications. We also need to think about refueling. For gasoline, the practice is on-board refueling—you drive into the gas station, pump some gas into the tank, and away you go. But for a hydrogen storage system, off-board refueling might be more likely. The charged hydrogen storage medium might be supplied in a cartridge. Refueling might involve swapping spent cartridges for full ones, and the spent ones will be recharged at a hydrogen station or some other purpose-designed facility.

So what are good candidates for hydrogen storage media? This is where the kind of chemistry I do becomes important. I am interested in main group chemistry. The main group elements are on the left and right sides of the periodic table (groups 1 and 2, and 13–18). Main group elements can be metals, nonmetals, or metalloids (elements with in-between characteristics). Importantly for our application, the lightest elements are main group elements, compared to metals like iron, cobalt, nickel, copper, zinc, silver, and gold, which are transition elements, and, being heavier, they are not as useful for hydrogen storage applications because of the weight percent targets required by the Department of Energy. Fossil fuels (hydrogen and carbon) could be hydrogen storage materials—they have high energy density but do not have low energy pathways for reversible removal of hydrogen.

What elements does the periodic table offer us in place of the carbon found in hydrocarbons? The elements directly to the left and right of carbon are boron and nitrogen. Simple valence electron counting shows us that boron (three electrons) combined with nitrogen (five electrons) has the same number of valence electrons as two carbon atoms (four electrons each). Take a simple hydrocarbon like ethane, $H_3CCH_3$, and imagine replacing the two carbon atoms by boron and nitrogen, which would result in an isoelectronic compound with the formula $H_3BNH_3$. This is not just hypothetical; it's a real compound called ammonia–borane and is made from borane ($BH_3$, which usually exists as a dimer, $B_2H_6$) with ammonia, $NH_3$. Ammonia–borane is an air-stable white solid containing 19.6% hydrogen by weight, which melts at around 110–115°C and is soluble in a range of common solvents. More importantly, ethane is a nonpolar molecule (there is little electronegativity difference between C and H), but the B–N bond in ammonia–borane is quite polar, and this opens up chemical reaction pathways for the removal of hydrogen.

Under the right chemical conditions, ammonia–borane loses hydrogen. The simplest reactions (on paper) are loss of one $H_2$ from $H_3BNH_3$ to give aminoborane ($H_2BNH_2$), loss of a second $H_2$ to give iminoborane (HBNH), and finally, loss of the last $H_2$ to give boron nitride, BN. Aminoborane and iminoborane are isoelectronic analogues of the hydrocarbons ethene (ethylene, $H_2CCH_2$) and ethyne (acetylene, HCCH). However, unlike ethane and ethyne, which are well-known gases, the B–N analogues can only be prepared and observed under very stringent conditions and are very chemically reactive. One other well-known example of a hydrocarbon analogue containing boron, nitrogen, and hydrogen is borazine, $B_3N_3H_6$, which has a structure very similar to benzene but contains alternating B and N atoms in place of the six carbon atoms that occur in benzene. Boron nitride, BN, is isoelectronic with pure carbon and, like carbon, can occur in both diamond and graphitelike forms (Table 15.1). The diamond form of BN is very hard and is used in abrasion-resistant devices, while the graphitic form has a high sheen and is used in cosmetics like lipstick and face powder to give them their luster.

**TABLE 15.1. Boron–Nitrogen Analogues of Simple Hydrocarbons**

| Carbon | Graphite Diamond | Boron nitride | Graphitic form Diamond form |
|---|---|---|---|
| Alkane $H_3CCH_3$ | | Ammonia–borane $H_3BNH_3$ | |
| Alkene $H_2CCH_2$ | | Aminoborane $H_2BNH_2$ | |
| Alkyne HCCH | H—C≡C—H | Borazyne HNBH | H—B≡N—H |
| Benzene $C_6H_6$ | | Borazine $B_3N_3H_6$ | |

These boron–nitrogen compounds are a good example of the relationship between fundamental chemistry and applied chemistry. My research in main group chemistry has centered around the reactions of compounds containing boron and nitrogen. My colleagues and I have asked ourselves why two molecules that have the exact same number of electrons, but one contains carbon atoms while the other contains boron–nitrogen pairs, have different chemical reactivities. We have studied molecules related to ammonia–borane, aminoborane, iminoborane, and borazine to try to answer this question and to understand more about what factors control the structure and reactivity of these and many other molecules. This can be described as fundamental chemistry—asking questions that help us to learn more about the underlying drivers in the chemical world. The search for new hydrogen storage media is a recent endeavor that draws on and develops the work of fundamental chemists for the purpose of solving an applied problem, how to find a safe and effective way to store hydrogen as a vehicle fuel.

Let's get back to what happens when hydrogen is removed from ammonia–borane and whether it might be useful as a hydrogen storage medium. What questions do we need to ask? Let's call it AB—it's less of a mouthful than ammonia–borane. How do we initiate removal of hydrogen from AB? What chemical compounds form when AB loses hydrogen? How much hydrogen can be extracted from AB? How can we add hydrogen back to AB so it can be recycled? This is all work that chemists need to do.

Since AB doesn't spontaneously lose hydrogen at room temperature, the process must be initiated, but remember that in order to be sustainable this process must not use too much energy, and if other chemicals are required, they can't be too heavy (the weight percent targets), expensive, toxic, nonrenewable, or nonrecyclable. That's a tall order and chemists need a big toolbox of knowledge and techniques—much of it developed by fundamental chemists. Hydrogen removal can be initiated by thermolysis (heating in the solid state or gas phase), by hydrolysis (addition of water), or it can be catalyzed by the addition of acid (a proton acid or a Lewis acid), a small amount of a metal, or a metal complex (Fe, Ni, Rh, or Ir complexes, calcium hydride), or on a solid surface (NiPt microspheres or nanoporous silica). All those potential systems need lots of work to understand their kinetics (reaction rates) and thermodynamics (energy balances) and to characterize the B-N-containing reaction products.

The optimum amount of hydrogen released from AB is about 2.2 molecules of $H_2$ for every molecule of AB. If all the hydrogen is removed to form boron nitride, then the process has gone too far downhill in energy terms and it will cost too much energy to reform AB—remember, the energy cost of every step is critical if AB is to be a viable storage medium. It turns out that removal of hydrogen from AB doesn't give the simple molecules aminoborane and iminoborane but instead forms a mixture of one-, two- or three-dimensional structures linked by new B–N bonds that form as hydrogen is removed. This material is the "spent fuel" we referred to above and needs to be processed to reform AB. For a vehicle application, the dehydrogenation needs to occur on-board; in other words, hydrogen gas needs to be released at the point where it will be used. But the reprocessing of the spent fuel can be done off-board, and so the requirements are not so stringent, although cost-effectiveness (energy and dollars), safety, and sustainability issues still exist.

The reaction of the AB spent fuel directly with $H_2$ to reform AB (direct hydrogenation) is not feasible, so the reprocessing might require several linked chemical cycles ideally using recyclable materials. Here's how it might work (Figure 15.3). The spent AB is digested with a sulfur reagent that breaks down the B–N bonds and replaces them with B–S bonds. The B–S compound is then treated with a metal hydride compound (e.g., containing tin and hydrogen) and ammonia, which reforms AB and produces a tin–sulfur product, which in turn is treated with hydrogen ($H_2$) to reform the sulfur reagent and the tin hydride. So the only material that is consumed is hydrogen (sourced from water), and the sulfur reagent and tin hydride

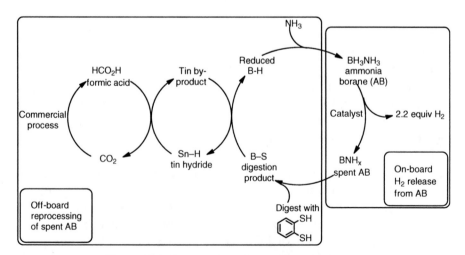

**Figure 15.3.** Regeneration of spent ammonia–borane.

are recycled. Sulfur and tin are both main group elements, and metal hydrides in general also have rich and varied chemistry—lots of work to keep my colleagues and our students busy, but with the promise of an exciting outcome when the solutions to these challenges all come together.

Here's how it all might work. You pull into a hydrogen station, exchange your 10-kg spent fuel cartridge for a new one full of fresh AB. The fuel cell in your car runs on oxygen from the air and hydrogen, which is supplied from a catalytic reaction releasing $H_2$ from the AB at the rate at which it is required by the fuel cell. Electricity from the fuel cells powers your quiet, pollution-free electric car and also charges the batteries required for ignition, fans, wipers, lights, and your iPod. When the $H_2$ supply from the AB cartridge runs low, the supply switches to the backup cartridge and you pull into a station to switch the old one for a new one. A nearby facility digests the spent AB and produces rehydrogenated AB using a continuous flow process in which the sulfur and tin reagents are continuously recycled. The hydrogen is produced by electrolysis of water using electricity from solar photovoltaic collectors, which form an integral part of the plant. Water is the only feedstock consumed by the plant, and oxygen (the other component produced from electrolysis of water) is the only by-product. We are now at the stage where the drive to the beach to start our camping vacation has eliminated the need for fossil fuels!

## SUNTAN WITH A PURPOSE: CONVERTING SUNLIGHT TO ELECTRICITY

The part of our sustainable camping trip that we haven't really looked at is the whole question of actually converting sunlight energy to a usable form—needed to produce the electricity to drive our water electrolysis, and also needed so that we have alternative sources of energy while we are actually camping. We are already familiar with chemical changes induced by sunlight—brightly colored fabric left in the sun will fade, and people with fair to moderate skin tones undergo changes in skin color (burning or tanning) when exposed to the sun. Photovoltaic devices convert sunlight directly to electricity by triggering a chemical change in the material absorbing the sunlight. This chemical change—how to induce it, how to collect the electrical current, how to understand which materials can be used, and how to make the process more efficient are all of critical concern to chemists, and we'll talk a little about this now. We'll look at the photovoltaic devices already in use, a new generation currently in development, and the futuristic ones that will really revolutionize how we utilize solar energy.

Silicon-based photovoltaics are already in widespread use—they power satellites in space and other applications requiring local generation of electricity, like radios in mountain refuges. They are scalable—the smallest units are individual cells, which are grouped together in modules, and these can be linked into solar photovoltaic arrays for large-scale generation of electricity from sunlight, which can feed into a power grid. The cells are silicon chips, and, like those used in the computer and electronics industries, are manufactured from high-purity silicon. A simple heterojunction cell comprises two layers of silicon, one p-type (doped with boron, which produces positively charged electron holes) and the other n-type (doped with phosphorus, which produces additional negatively charged electrons). When the cell absorbs a photon of sunlight of the right energy, an electron is excited from the valence band (where it is tied up in electron pair bonds) into the conduction band (where it is free to move through the bulk material). This creates an electron–hole pair, which, when the electron and hole are drawn off in opposite directions, creates an electrical current (Figure 15.4). The challenge is to keep the electron and hole separated for long enough for the current to flow. The alternative is recombination, which occurs when the electron and hole recombine and no current flows. The electric field that occurs at the heterojunction where the layers of p-type and n-type

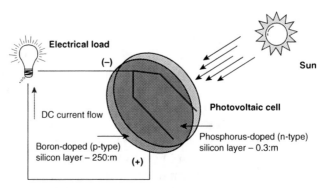

**Figure 15.4.** *Silicon photovoltaic cell. Image from http://www.blueplanet-energy.com/images/ solar/PV-how_it_works.gif.*

silicon meet is critical in keeping the electron and hole separated for long enough for the photocurrent to flow. The typical efficiency of these kinds of cells is around 10%, which means that only 10% of the light energy is converted to electricity. Losses in efficiency occur in the absorption step (light is reflected from the cell or passes through without being absorbed) and through premature recombination of the electron–hole pair. Silicon has a low band gap, which means that the energy gap between the valence and conduction bands is low. Photons with energy lower than the band gap can't be absorbed, whereas photons with energy higher than the band gap can be absorbed, but the excess energy from these photons is lost as heat, another way in which the efficiency is reduced.

The advantages of silicon photovoltaics are that the technology (developed in tandem with the electronics industry) is well established and different forms of silicon (e.g., amorphous silicon) are being developed as alternatives to ultrapure crystalline silicon. The low efficiency and rigid, fragile nature of the modules limits their uses. There's also a direct correlation between the surface area of the cells and the amount of energy that can be captured. For example, a typical domestic car doesn't have a high enough surface area on its roof, hood, and trunk to collect enough energy to run the car.

How can we increase the efficiency of the cells? One possibility is to explore different semiconductor materials with different band gaps—if there's a better match between the band gap and the incoming solar radiation, then more of the energy can be converted to electricity. This is where we come back to some of the ideas we explored when we were looking at main group elements. Silicon is in group 14, directly below carbon. Remember when I talked about boron–nitrogen compounds as materials for hydrogen storage, and I pointed out that a boron–nitrogen pair of atoms has the same number of valence electrons as two carbon atoms. More generally, any group 13–group 15 pair of atoms will have the same number of valence electrons as two group 14 atoms. Boron and gallium are in group 13, and nitrogen and arsenic are group 15 elements. So gallium arsenide (GaAs) is a semiconductor with many similarities to silicon but with a different band gap—it can absorb higher-energy photons. Multijunction cells containing GaAs and other group 13/15 semiconductors have achieved efficiency of up to 35%. As inorganic chemists, we have to ask how GaAs (and a whole range of other group 13/15 semiconductors) can be prepared using, of course, the principles of green chemistry so that the whole process is sustainable. This is another flourishing area of main group chemistry that I'm interested in, making compounds that can be used as precursors to group 13/15 semiconductors. For example, $(CH_3)_3GaAsH_3$ is related to ammoniaborane, $H_3BNH_3$, by conceptually replacing B by Ga, N by As, and H by $CH_3$. GaAs could be made from $(CH_3)_3GaAsH_3$ by pyrolysis, heating it up to remove the carbon and hydrogen as methane, $CH_4$.

In these kinds of photovoltaic cells, the absorption of energy from sunlight and charge separation into the electron–hole pair both occur within the semiconductor material. In a new generation of cells currently under development, these two processes occur in different materials. A semiconductor is still used, but this time it is titanium dioxide ($TiO_2$), which is a high band gap material. This means that only a small part of the solar spectrum has enough energy to excite an electron, so $TiO_2$ on its own isn't very useful. Chemists have solved this problem by coating the $TiO_2$ with a dye that will absorb a large proportion of the solar spectrum. The dye is carefully chosen so that when it absorbs light, an electron in the dye is excited into a higher-energy state, and this excited electron is then transferred to the semiconductor (a process called injection). In this way, the free electron resides in the $TiO_2$, while the hole resides in the dye molecule. With both the semiconductor and the solution containing the dye in contact with electrodes, the electron–hole pair can be drawn off as electrical current. These kinds of cell are referred to as dye-sensitized solar cells (DSSCs) (Figure 15.5). They have lots of potential advantages— the cells are more robust, and $TiO_2$ is cheap, abundant, and can be used as an amorphous (noncrystalline) coating on a suitable substrate. There are a huge range of possibilities for the absorbing dyes— many of them are metal complexes, and some that contain porphyrins (found in hemoglobin, which gives blood its red color, and related to chlorophyll, which gives leaves their green color) are being studied, all offering more exciting possibilities for chemists. DSSCs still need further improvement, as their efficiency is similar to that of silicon-based photovoltaics (around 10%). In addition, they usually contain a solution or gel as the electrolyte, which can reduce their lifetime and limit the kinds of materials that can be used to construct the cells.

We have looked briefly at the silicon photovoltaic devices already in use and the new generation of DSSCs currently in development. What if we take a leap in imagination and rather than design solar *cells* that need to be deployed to catch the sun (the solar collector on top of your roof), we think about designing solar *materials* that can do two jobs at once (the tiles on your roof themselves actually

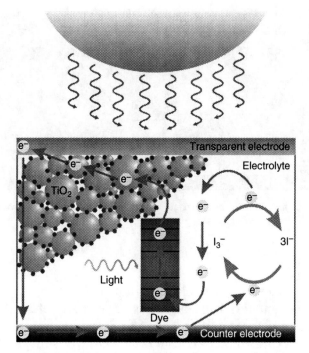

**Figure 15.5.** Dye-sensitized solar cell. Image from http://commons.wikimedia.org/wiki/ File:Dye_Sensitized_Solar_Cell_Scheme.png.

become the solar collectors). For a photovoltaic device, we still need the two essential steps, absorption of sunlight and formation of an electron–hole pair. If these processes can occur in materials that are already in use, then we have multiple advantages. Surfaces that already catch high amounts of sunlight could be converting it to electricity—think of the roof of your house, the paint on top of your car, road surfaces, and, to get back to our sustainable camping notion, the fabric of your tent.

What issues will we need to address? We will need to marry photovoltaics and electronics with the plastic and polymer fields to learn how to use the techniques of polymer chemistry, which gives us flexible, durable films, coatings, paints, and fabrics that can be modified to capture solar energy and convert it to electricity. There is already a thriving new research area of conducting polymers (in 2000 the chemistry Nobel Prize was awarded for this)—materials through which countercurrents of electrons and holes can flow. If a conducting polymer can be impregnated with an absorber that can use incoming

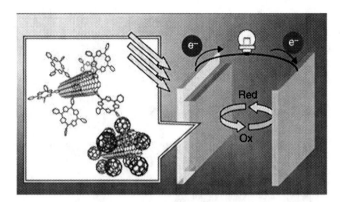

**Figure 15.6.** *Organic solar cell. Image from http://www.rsc.org/Publishing/ChemTech/ Volume/2008/07/Imahori_insight.asp.*

solar energy to generate the electron–hole pairs, then we would be on the way to new materials. These kinds of cells are referred to as bulk heterojunction devices and comprise an interpenetrating network of donor and acceptor materials. The donor absorbs sunlight and generates a high-energy electron, which is transferred to the acceptor. There are all kinds of problems to solve (problems are good things to chemists, something to get excited about and which open doors to new ideas) to do with suitable donor and acceptor materials, how to get them aligned on a "nano" scale for efficient charge carrying, how to prevent a recombination of the electron and hole, how to achieve the larger-scale material characteristics that will make them usable as fabrics, paints, roof tiles, or road surfacing materials. This work also leads to surprising connections within chemistry—for example, one of the molecules being investigated as an electron acceptor in polymer devices is "buckyball," the $C_{60}$ fullerene (1996 Nobel Prize in chemistry), which has the same symmetry as a soccer ball or Buckminster Fuller's geodesic domes (Figure 15.6).

So coming back to the camping trip where we started, our futuristic tent will be made from lightweight, strong, waterproof material that will also have a polymer coating enabling it to capture and convert sunlight to electricity—only on a small scale, but our camping needs are modest. This electricity will be used to charge a new generation of lightweight, portable batteries so that the energy collected on sunny days is available on cloudy days and at night. We could use this for camp lights, to run a small cooler for our sodas, and to recharge our iPods.

## BRINGING IT ALL TOGETHER

We've looked at ideas to do with energy generation and storage just by beginning with the very simple notion of a family camping trip. Hopefully, it illustrates that once we start to think sustainably, we have to analyze every single item we use and every move we make in order to truly understand what changes will be required. But the answers to the big sustainability challenges will be high tech and will require us to marshal all of our skills and knowledge across a whole range of disciplines. Most of all, chemists will be integrally involved in all of this—many new ideas have been created, and research to make them a reality is underway. But even more new ideas are out there waiting for a new generation of dreamers who will also be the new generation of chemists who will make them happen—maybe one of them will be you!

Best wishes,

Penny Brothers

## FURTHER READING

Ashley, S. On the road to fuel-cell cars. *Scientific American* 2005, 62–69.

Birch, H. The artificial leaf. *Chemistry World* 2009, 42–45.

Grätzel, M. Recent advances in sensitized mesoscopic solar cells. *Accounts of Chemical Research* 2009, *42*, 1788–1798.

Hamilton, C. W.; Baker, R. T.; Staubitzc, A.; Manners, I. B–N compounds for chemical hydrogen storage. *Chemical Society Reviews* 2009, *38*, 279–293.

Mayer, A. C.; Scully, S. R.; Hardin, B. E.; Rowell, M. W.; McGehee, M. D. Polymer-based solar cells. *Materials Today* 2007, *10*, 28–33.

Satyapal, S.; Petrovic, J.; Thomas, G. Gassing up with hydrogen. *Scientific American* 2007, 81–87.

# 16

## *Clean Electrons and Molecules Will Save the World*

**Carl C. Wamser**
*Portland State University*

Carl Wamser is professor of chemistry at Portland State University (PSU). He earned an ScB degree at Brown University, doing his initial research activities in the group of Joseph Bunnett on organic reaction mechanisms. He received his PhD at Caltech, where he worked with George Hammond in the area of organic photochemistry. A postdoctoral year at Harvard University with P. D. Bartlett was followed by his first faculty position in 1970 at California State University, Fullerton. Research in the Wamser group at PSU is currently focused on solar energy conversion using porphyrins and conductive polymers.

*Letters to a Young Chemist*, First Edition. Edited by Abhik Ghosh.
© 2011 John Wiley & Sons, Inc. Published 2011 by John Wiley & Sons, Inc.

Hi Angela,

Your mom told me recently that you're considering chemistry as a major, and she thought I might have some suggestions or advice for you. Well, of course I do! Chemistry, as you know, runs in our family, certainly the Wamser branch of it. I got thoroughly dazzled by the wonders of chemistry from a very early age. My dad loved to show off various chemistry tricks to kids. Your mom would have been one of the "cousins by the dozens" who got to see the show during family visits. Ask her what she remembers best about those—maybe it was the color changes (pour the same colorless liquid into three different colorless liquids, and they go red, white, and blue!) or the clock reaction (pour one colorless liquid into another colorless liquid and ... nothing happens! ... until dad—Uncle Chris—magically taps the flask and it goes ink black). I got to see many more over the years, and I eventually learned to figure them out and to do my own experiments, all of which was very satisfying. But what I really learned from my dad is that there's nothing as important as loving what you do. You're going to spend a good fraction of your life with your career—make it fit with what you really love to do, and it will be rewarding and fun.

My dad is now 97 years old and still loves and stays close to chemistry. He tells me what I should read in the current journals (they usually get to the East Coast about 2 days ahead of the West Coast, so he's always got the jump on me). It's hard to imagine anyone holding onto a career so long and so lovingly—but chemistry can do that to you. I hope you find it that engaging and long-lasting. Let me tell you why I enjoy it so much.

## WORK ON SOMETHING IMPORTANT

I usually start any seminar I give with a slide taken (with permission) from Rick Smalley. Smalley won the Nobel Prize in 1996 for his part in discovering $C_{60}$, Buckminsterfullerene, a new form of carbon wrapped up into a soccer ball sphere. But in the final years of his life, Smalley was on a clear mission to alert the public and the scientific community to the urgency of key issues facing the world today. He'd ask the audience to identify the top 10 issues that should be addressed to improve the quality of life—globally—before the middle of this century. He claimed that any good audience would pretty much always come up with about the same 10 issues—and I've found that to be true too (at least 8 or 9 of the same issues always show up, sometimes in slightly different versions). If you want to try this yourself, don't peek at the answer key below.

In no particular order, the list includes the following. Note that most of these issues can be presented in either a negative (disease) or a positive (health) frame.

The top 10 issues to address before 2050 (R. Smalley) are

- energy
- water
- food
- environment
- poverty
- terrorism/war
- disease
- education
- democracy, and
- population.

The time frame (40 years) is an important point. Forty years from now, you'll be looking back on your career and asking yourself "Did I make a difference?" If you want to make a difference, you should address your efforts and talents to something important. The issues above are the most important that we can think of. Smalley would then make the point that if we could solve the energy issue, it would have a positive effect on a remarkable number of other issues on the list. If you can imagine cheap, abundant, and clean energy available worldwide, then it would be possible to create clean water from the sea by desalination (currently much too expensive because of the energy required), to irrigate the desert to grow food, to avoid mining and drilling for fossil fuels that lead to environmental degradation and global warming, and to avoid international tensions (and worse) associated with maintaining stable supplies of fuels. Energy is wealth; if you have enough energy, you can do what needs to be done.

So if you can't quite decide what you want to work on, let me humbly suggest taking something from this list. I've chosen to devote my career to education and to energy (solar in particular). And I frankly feel really good about the choices, even if I dearly wish that I and the rest of the world were making much faster progress.

## ENERGY

What makes energy such an important issue is its centrality; it enables everything we do. What makes it such a difficult issue is the sheer magnitude of the amount of energy we have come to expect in our daily lives. If you could plug the world into an energy meter, we'd be running an average of about 15 TW of total power continuously—a little more when the United States is awake and a little less when we are asleep. Although the United States is about 5% of the world's population, we use about 25% of the world's energy. So Nate Lewis (Caltech professor of chemistry) calls this the "terawatt challenge." (In case you haven't learned all the prefixes that high, these are the same units you hear used for computers—in multiples of 1000, they go kilo-, mega-, giga-, and tera-. So, a terawatt is $10^{12}$ W.) The world uses about 15 TW today, and that's estimated to double by 2050 and to triple by 2100. Since 85% of the energy we use today is from fossil fuels, continuing with that same mix of resources would (1) require an immense increase in the extraction of fossil fuels, and (2) drastically add to the atmospheric burden of carbon dioxide and global warming. We just can't do that. We need huge, new sources of energy (terawatts) that do not involve fossil fuels. Carbon-free energy resources is the way it's usually said.

A while ago, I attended a lecture by Thomas Friedman (*New York Times* columnist who primarily focuses on economic issues). After drawing up a daunting list of the crucial problems facing the world (not unlike our top 10 list, but he just picked five), he pointed out that a single solution could fix them all. What the world really needs is … clean electrons and molecules! I practically jumped out of my seat (which just showed that I hadn't read his book yet, since he used those terms there—in *Hot, Flat, and Crowded*). But what the world needs now is (well, love, of course) electrons and molecules! That's chemistry—chemistry is just making electrons move and molecules change into other molecules. Clean electrons means, for example, photovoltaics—using sunlight to create electricity. Clean molecules means generation of fuels from renewable sources, such as solar energy. Since Thomas Friedman had cited Nate Lewis a couple of times in his talk, I asked Nate if he had put him up to the "clean molecules" part. He had; Friedman had been calling for clean electrons, but Nate pointed out that the sun doesn't shine all the time, and you really do need storable and transportable energy (fuels, a.k.a. "molecules") to cover your energy needs all the time.

Although you can extract energy from wind or waves or running water, the most concentrated forms of energy come with chemical bonds. That's why you can use the energy in a gallon of simple hydrocarbons to drive a ton of metal 20 mi (or 45 if you're in my Prius). That's why Mother Nature does photosynthesis and why we get our daily energy fix from food. It's all chemistry, and it's an integral part of understanding and planning our energy future. Numerous scientists are working on ways of storing energy in chemical bonds starting from simple renewable resources. The simplest is splitting water to hydrogen and oxygen, using the energy of sunlight. Some chemists have called that reaction the "Holy Grail" of chemistry: sunlight and water making hydrogen and oxygen. It's deceptively simple but still difficult to carry out efficiently. Other reactions include converting cellulose (corn cobs seem to be the favorite lately) to ethanol or reducing carbon dioxide to useful products (e.g., methanol). I find it inspiring to know that Caltech, MIT, and several other institutions are collaborating to find a solution to just this problem. They call the joint project "Powering the Planet" and are currently working on solar water splitting.

## SOLAR ENERGY

Not only does the world need to add another 15 TW of carbon-free energy resources in the next 50 years, but if we hope to keep the atmospheric carbon dioxide from growing to extraordinary levels, then much of the current 15 TW of energy resources needs to be transformed into carbon-free forms as well. This is asking a lot; estimates are that we need 20–25 TW of carbon-free energy resources to stabilize carbon dioxide levels in the atmosphere, much less to bring them back to preindustrial levels. Where can we find such huge resources? There are sufficient coal reserves to carry us for another century or two, but of course, coal is not carbon free (in fact, it's pretty close to pure carbon). Estimates have been made recently to evaluate the total ongoing energy available from various resources: wind energy is about 2–4 TW; geothermal energy could be up to 12 TW; hydroelectric energy is mostly tapped out but might have another 1 TW available; and nuclear power plants are typically built in about 1-GW sizes. Since it takes 1000 GW to make 1 TW, we'd have to build 1000 1-GW nuclear plants (three a day for a year) to get our first TW.

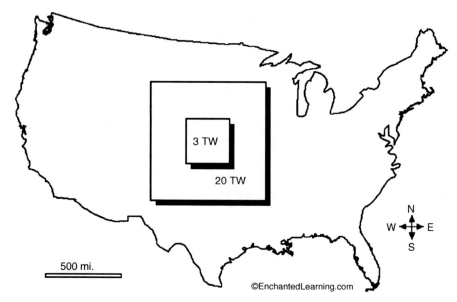

**Figure 16.1.** *Approximate area needed for solar energy (at 10% conversion efficiency) to provide about 3 TW (current U.S. consumption of all forms of energy) or 20 TW (approximately what's needed globally in carbon-free energy by 2050).*

What about solar energy? Earth is constantly bathed in 120,000 TW of solar energy! Without the sun, we'd be a frozen rock flying through space. The total amount of energy consumed by humans in a year arrives in just 1 hour of sunlight. In order to generate the 3 TW currently used in the United States, only a relatively small area is needed. Figure 16.1 is modified from one that Nate Lewis uses; it points out the area that would be needed, allowing for a modest 10% conversion efficiency, to supply all of the U.S. energy needs (about 3 TW currently) or to supply what the world will need in 2050 (about 20 TW). Even the large square in the figure is smaller than the land area we've already dedicated to growing corn and other crops, which really just amounts to collecting solar energy for food. We have the space and we have enough solar energy; we just need the dedication to get it done.

## PHOTOSYNTHESIS

Nature has created a wonderful model for how to convert solar energy into useful products. Photosynthesis has evolved over millions of years into a beautifully complex mechanism for converting carbon dioxide and water into carbohydrates plus oxygen:

$$CO_2 + H_2O \xrightarrow{\text{sunlight}} (CH_2O) + O_2.$$

Photosynthesis is a magnificent reaction. It takes carbon dioxide out of the atmosphere (that's especially good these days), adds water and sunlight (both plentiful and cheap), and creates carbohydrates (food, wood, and everything else we get from plants) plus oxygen (good breathable stuff). The detailed mechanism for accomplishing this, however, is incredibly complex. We're getting to know more and more about the key issues that make it work so well. Chief among these are the need for efficient light absorption (chlorophyll does that) and a carefully arranged alignment of molecules to set up an electron transport chain. When everything is properly arranged, absorption of light leads rapidly to charge separation. The speed and the efficiency are amazing. Essentially, every photon absorbed leads to electron transfer in picoseconds. (You also need to know all the prefixes in the other direction: as you divide by 1000, it's milli-, micro-, nano-, pico-, femto-, atto-, zepto-. ...) So, picoseconds are $10^{-12}$ seconds, barely enough time for a molecule to vibrate but enough time for an electron to move (a little).

A variety of approaches, often called artificial photosynthesis, borrow either the molecules of nature or the strategies of nature. Some incredibly functional molecular systems have been created that mimic many of the initial steps of photosynthesis, most notably from the group at Arizona State University (Professors Devens Gust, Thomas Moore, and Ana Moore). For example, a porphyrin, which is a structural analogue of chlorophyll, can be covalently connected to an electron donor on one side, such as a carotene, and an electron acceptor on the other side, such as a quinone (Figure 16.2).

Light absorption by the porphyrin energizes an electron (creating an excited state, S*); the excited electron moves to the quinone, and the vacancy (often called a hole) is filled by moving an electron in from the carotene. The porphyrin is back where it started, ready to absorb another photon of light, and we've created oxidized and reduced forms separated on either side. This is completely analogous to what happens in the photosynthetic membrane, where light absorption by chlorophylls ultimately leads to reduced quinones on one side of the membrane (where eventually $NADP^+$ gets reduced to NADPH, and that goes on to reduce $CO_2$ to carbohydrates) and to oxidized molecules on the other side of the membrane (where eventually

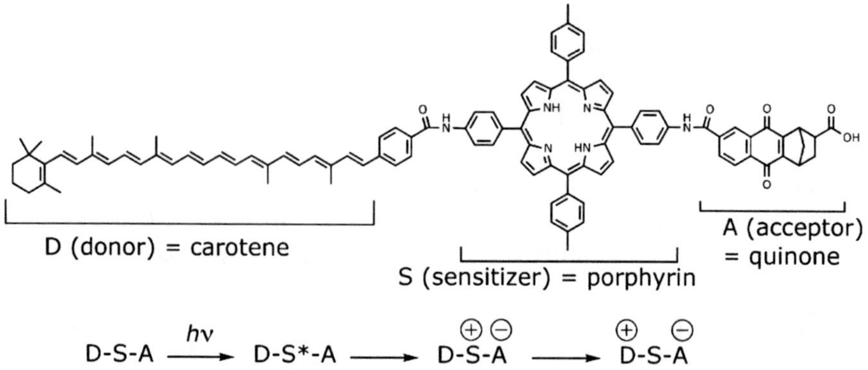

D (donor) = carotene

S (sensitizer) = porphyrin

A (acceptor) = quinone

$$\text{D-S-A} \xrightarrow{h\nu} \text{D-S*-A} \longrightarrow \overset{\oplus}{\text{D}}\text{-}\overset{\ominus}{\text{S}}\text{-A} \longrightarrow \overset{\oplus}{\text{D}}\text{-S-}\overset{\ominus}{\text{A}}$$

**Figure 16.2.** *A three-component molecule (called a triad), stimulated to undergo charge separation by the energy in a photon of light, represented by hν.*

water gets oxidized to $O_2$). This basic concept has been extended to larger molecules, tetrads and pentads, with the charge separation even further extended in both distance and time.

## SILICON SOLAR CELLS

Most solar cells are based on silicon. This has been true pretty much for the whole history of solar cells (over 50 years). Silicon is a tough competitor, primarily because it has an almost ideal absorption spectrum relative to sunlight. Sunlight comes in a wide range of wavelengths—from about 300 nm in the ultraviolet to very long wavelengths in the infrared. But our eyes can only see about 400–800 nm (the visible spectrum). Silicon absorbs everything out to nearly 1100 nm, into the near infrared. This turns out to be an ideal absorption range, given the range of wavelengths in sunlight. When planning for solar energy conversion, efficient harvesting of the light is the first important factor to take into account. Although silicon absorbs all wavelengths up to 1100 nm, it ignores photons of higher wavelength, where nearly half of the total solar energy is contained. But those photons just don't have enough energy to energize electrons—infrared photons typically cause vibrations (heat) but can't generate electronically excited states. The lowest energy that can cause electronic excitation in a molecule or material is called its band gap—the energy spacing between the highest filled orbitals (or valence band) and the lowest available unfilled orbitals (or conduction band). So silicon's band gap is about 1100 nm or 1.1 eV.

A second limitation on light harvesting is the way that high-energy photons are utilized. For photons with energy greater than the band gap, the excess energy is degraded to vibrational energy or heat. Thus, all photons, even in the ultraviolet with very high energy, act no better than photons at the band gap. By the time you're done degrading the high-energy photons and missing the low-energy photons (beyond the band gap), the best you can do is convert about 30% of the total solar energy into electrical energy. The optimum harvesting occurs with a material having a band gap about 1000 nm. Silicon is close to ideal—the best silicon solar cells can get up to 20% efficiency. Nature uses chlorophyll, which absorbs out to 700 nm, and does very well with that. Most synthetic materials being explored for solar cells manage to harvest efficiently up to about 700 nm but have trouble going beyond that. Silicon is indeed a tough competitor.

Silicon is also versatile. In a silicon cell, the same material is used for both the light absorption and charge transport. Silicon is doped with small but very precise amounts of impurities that render it either n-type (favoring conduction of negative charges) or p-type (favoring conduction of positive charges). When silicon absorbs a photon, electrons prefer to move to the n-side and holes move to the p-side. The result is a photocurrent (electron flow), which can be up to 30 mA/$cm^2$ in full sunlight, and a photovoltage, which can be up to 0.6 V. Although the band gap is 1.1 V, the extracted voltage is substantially lower because of the energy difference created at the p–n junction and internal losses (Figure 16.3).

Despite the impressive history and accomplishments, silicon has some weaknesses. The majority of solar cells are currently made by slicing thin wafers from cylinders of highly purified, single-crystal silicon. This is very expensive, energy intensive, and ultimately

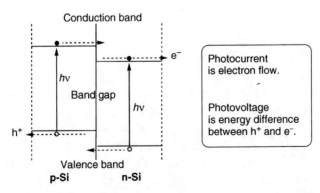

**Figure 16.3.** *Energy levels and electron/hole movements in a silicon solar cell.*

wasteful (40% of a typical cylinder is lost as "kerf"—sawdust from slicing up the cylinder and polishing the wafer). If we intend to greatly expand the use of solar energy, we need cells made from cheaper materials that can be processed simply. Thin-film solar cells use amorphous silicon, avoiding the need for generating huge single crystals. But the efficiency is lower because mobility of charges through the amorphous material is less efficient than mobility through a single crystal. Overall, solar energy conversion efficiency is about 5–8% for amorphous Si versus 12–18% for monocrystalline Si.

## ORGANIC SOLAR CELLS

Organic molecules that absorb light are called dyes. Whenever a molecule absorbs visible light, we can see the effects; a dye that absorbs blue light tends to look red, and a dye that absorbs red light tends to look blue. We see the portion of the spectrum that is not absorbed; it's reflected back and detected by our eyes. Clearly, Mother Nature has chosen green as the primary color for solar energy; chlorophyll absorbs both the blue and red ends of the visible spectrum, leaving only a sliver of green unused in the middle.

The first fully organic solar cell was prepared in 1986 by C. W. Tang, who worked at Kodak, where they really know all about light and color, of course. He used two different dyes that absorbed different parts of the solar spectrum and had different energy levels (not unlike p- and n-silicon). The two dyes were layered in thin films such that either one could absorb the sunlight (Figure 16.4).

The required steps for this type of cell include (1) light absorption by either of the two dyes, (2) getting the excited state to the interface—the only place where donor and acceptor molecules are in immediate contact with one another, (3) charge separation at the interface by moving electrons into the acceptor layer and holes into the donor layer, and finally (4) moving those charges in opposite directions out to the collecting electrodes. There are serious difficulties with all of these steps, primarily because organic materials behave fundamentally differently from silicon. When an electron is excited in silicon, it moves directly into the conduction band and is mobile as a free electron. It leaves behind a hole, which is also freely mobile. An excited state in an organic molecule does not readily lose an electron to a neighbor—the energy required to generate an oxidized and a reduced molecule as nearest neighbors is generally too high when the

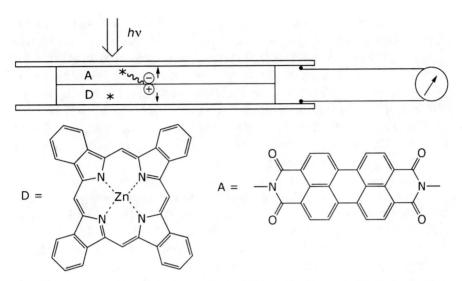

**Figure 16.4.** A bilayer organic solar cell—the original version by C. W. Tang used zinc phthalocyanine as the donor (D) and a perylene dye as the acceptor (A).

neighbors are identical molecules. Thus, the excited state can be considered to be an electron–hole pair bound together on the same molecule, called an exciton. Fortunately, the exciton is mobile. Nature is quite aware of this; hundreds of chlorophyll molecules are used as antennas, just to absorb light and to pass the exciton around until it arrives at the "special pair" of chlorophylls that are properly set up for charge transfer.

So an excited state, or exciton, in an organic solar cell needs to wander about, by random diffusion, until it arrives at the interface—only there does it find a nearest neighbor that is different from itself. The pair of molecules (D and A) is specifically chosen so that electron transfer will be favorable. Regardless of whether it was a donor exciton or an acceptor exciton that arrived at the interface, an electron will move into the acceptor layer and a hole will move into the donor layer. Now, however, we have to move the charges through the full thickness of each layer in order to get them to the collecting electrodes, and organic materials are generally not good conductors of electrical charges. So what's needed in this type of solar cell are two contradictory features: good light absorption calls for thicker films, but good charge collection needs thinner films. It's not surprising that the original Tang cell had an efficiency less than 0.5%.

Improvements in organic solar cells have taken two routes, addressing those two conflicting requirements. One is to roughen the interface so more molecules can be nearer their opposite counterpart. In the optimum case, what can be formed is a blend of the two materials (D and A) such that there is always a connectivity between all the Ds and all the As, yet they penetrate into one another's territory extensively. This is called a bulk heterojunction cell, and it allows for relatively thick films and good light absorption, yet no exciton ever has to diffuse very far to find a counterpart for charge transfer. The second essential strategy is to improve the electronic conductivity of the organic materials.

Organic materials that conduct electricity represented an entirely new concept. In general, organic materials are plastics, suitable for the insulation around a wire, not the wire itself. The Nobel Prize in 2000 recognized the work of Alan Heeger, Alan MacDiarmid, and Hideki Shirakawa in showing that appropriately constructed organic materials could be excellent conductors of electricity. What's needed for conductivity is extensive delocalization of molecular orbitals coupled with some doping—intentionally created extra electrons or holes that can be readily moved around the extensive delocalized network. Polyaniline represents a classic example (Figure 16.5).

When electrons are removed from the nitrogen lone pairs in polyaniline, the remaining holes (oxidized nitrogens with positive charges) are remarkably mobile. Optimum conductivity occurs when exactly half the nitrogens are oxidized. Conductivity diminishes if the polyaniline is reduced or overoxidized, or if the nitrogens lose their pro-

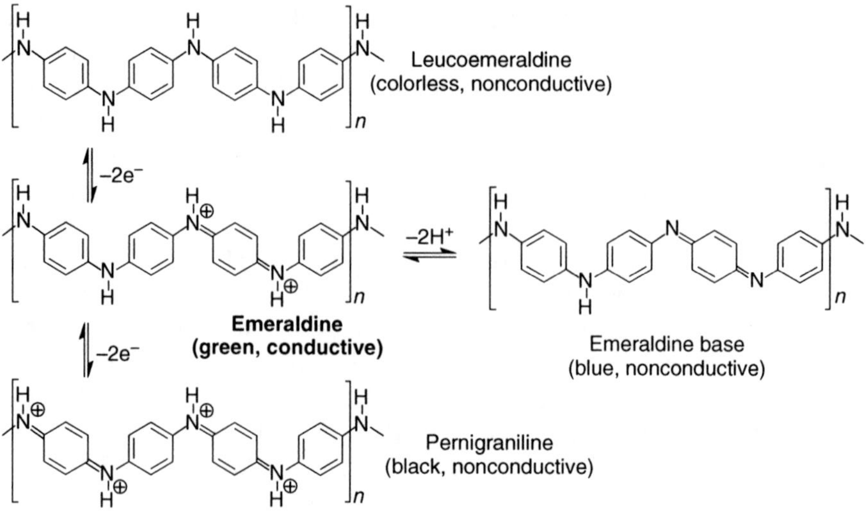

**Figure 16.5.** Structures and properties of polyaniline.

tonation (which makes the nitrogens nonequivalent and fixes the holes in place). Similar conductive organic polymers, most notably polythiophenes, are now regularly used in solar cells. Purely organic solar cells are starting to make their mark, primarily because they hold the promise of simpler and cheaper processing; efficiencies have exceeded 5%, and improved yields are reported regularly. A few companies are beginning to market organic solar cells (Konarka, Plextronics, and Risø).

## DYE-SENSITIZED SOLAR CELLS

The typical dye-sensitized solar cell is a hybrid type of cell, using both an inorganic semiconductor and an organic dye. In a dye-sensitized solar cell, all three primary functions are allocated separately, allowing for the possibility of optimizing each function independently. A dye is chosen for light absorption; a semiconductor is used to transport electrons; and a third component is used to transport holes. Currently, the most popular electron transport medium in dye-sensitized solar cells is titanium dioxide ($TiO_2$). $TiO_2$ is a wide–band-gap semiconductor, which just means that it does not absorb at all in the visible range, only in the ultraviolet. $TiO_2$ is a common white pigment for paints. (Pigments are insoluble particulate coloring materials, while dyes are soluble coloring materials.) $TiO_2$ is also found in toothpaste and a variety of other commercial materials; it is cheap and innocuous. When $TiO_2$ absorbs sunlight, it is capable of a good photovoltaic effect, but it only absorbs ultraviolet, which is less than 5% of sunlight. Nevertheless, if a dye can be tightly adsorbed on the surface of $TiO_2$, and the energy of the dye excited state is suitable to move an electron into the conduction band of $TiO_2$, then a photovoltaic effect can be initiated by all the photons that the dye absorbs. This is called dye sensitization, and it vastly extends the range of photons that can be harvested.

The next crucial component is the hole transport medium, which must replenish electrons to the oxidized dye. Again the energy levels must be appropriate for efficient electron (or hole) transfer. The analogy with the triad of artificial photosynthesis should be clear here. The dye is like the porphyrin (actually it often is a porphyrin) set in the middle of a donor and acceptor. After a photon creates an excited state, electrons are pushed one way (toward the acceptor or electron transport medium) and are pulled from the other side (typically, this is considered as holes transported away rather than electrons pulled in).

The strong point of the dye-sensitized solar cell is its geometric arrangement. $TiO_2$ is formed as a thin film of nanoparticles, like a thin layer of sand but in extremely small grains. In our lab, we make thin films of $TiO_2$ nanoparticles by spreading a thin slurry or paste of wet particles across an electrode that has transparent (plain Scotch) tape on two sides. The tape makes a boundary that limits the thickness of the paste as it's spread to about 50 μm. After removing the tape and drying the solvent, the nanoparticles are heated just enough to bind them together (sintering) without melting them or losing their incredibly high surface area. A typical cell for testing in the lab is $1 cm^2$ (and after sintering, about 15 μm thick). That $1 cm^2$ can have an internal surface area up to $1000 cm^2$! That surface is extremely active and can adsorb dye molecules onto the available surface sites of $TiO_2$. If you select the right functional groups in your dye molecule (carboxylic acids seem to work best), a complete monolayer of dye can be coated onto the surface of $TiO_2$. With such a huge surface area, a layer of $TiO_2$ nanoparticles only 15 μm thick can adsorb enough dye in just one molecular layer to absorb most of the solar spectrum. And every dye molecule is directly connected to the $TiO_2$ network so that the excited electron is immediately available to move into the electron transport medium ($TiO_2$ is a very good n-type semiconductor). Using a ruthenium-based dye with carboxyl anchors, Brian O'Regan and Michael Grätzel reported almost 11% solar conversion efficiency in the first report about this type of cell. That was in 1991, and efficiencies have only slightly improved since then; they got just about everything right the first time. Light harvesting by the dye is excellent, and electron collection through the $TiO_2$ is very efficient (Figure 16.6).

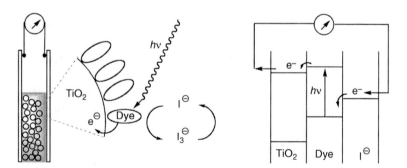

**Figure 16.6.** *Essential features of a dye-sensitized solar cell.*

I haven't yet mentioned the hole transport medium, and that, in fact, is the weak link in the dye-sensitized solar cell. The only thing that works well right now is to use a solution containing iodide. Iodide is a mild reducing agent, suitable for donating an electron back to the oxidized dye. Keeping the whole nanoparticle dye film soaked in an iodide solution allows iodide ions access to all the surface. As iodide gets oxidized itself, it is transformed to triiodide ($3\,I^- \rightarrow I_3^- + 2\,e^-$), which can diffuse to the counter electrode and regain electrons to regenerate iodide ($I_3^- + 2\,e^- \rightarrow 3\,I^-$). Thus, iodide serves as an electron relay via its redox reactions; it drops off electrons to oxidized dyes as needed, then picks up more electrons at the counter electrode. The net result is electrons pumped into the working electrode (from the $TiO_2$ conduction band) and pulled out at the counter electrode (into the iodide/triiodide redox couple).

The photocurrent from such a cell corresponds to the number of electrons flowing in the external circuit. In a good cell, this can be as high as $20\,mA/cm^2$ under illumination of standard sunlight. The photovoltage from a dye-sensitized solar cell is roughly the difference between the energy of the electrons coming into the working electrode (the energy level of the $TiO_2$ conduction band) and the energy of the counter electrode (in equilibrium with the iodide/triiodide redox couple). Photovoltages can be up to about $0.7\,V$ if there are minimal losses elsewhere in the system. Overall, these values lead to efficiencies of about 10%.

The primary advantage of dye-sensitized solar cells is in the simplicity and inexpensiveness of their construction. All materials are used in small quantities without the stringent purity requirements of silicon cells. Currently, one company (DyeSol) has begun marketing dye-sensitized solar cells. Nevertheless, it is clear that the glaring weakness is the need for a liquid phase to allow for the diffusion of iodide within the pores of the nanoparticles. Numerous efforts have been undertaken over many years to create a solid-state version of the dye-sensitized solar cell, but nothing has yet made a significant improvement on the original.

## SOLAR PHOTOCHEMISTRY

All of the above is completely fascinating, of course. Yet so far, we've really only considered pushing electrons around—very little real chemistry, if you think chemistry means changing molecules into different molecules. What about using solar energy to do real chemistry? That's what nature does in photosynthesis. Making real molecules from just $CO_2$ and $H_2O$ is very impressive chemistry. Too impressive for humble chemists to emulate exactly, but we do dream about it. So from this point, you're only limited by your imagination. What useful reactions should we try to create, using sunlight as the energy source?

One approach is to develop simple reactions, starting with abundant materials and ending up with high-energy products. Splitting water to hydrogen and oxygen is the classic example. But you could also visualize reactions that reduce carbon dioxide to various other forms. To be truly useful, the reactions should gather a significant amount of energy (from sunlight) and generate a fuel that can readily return that energy upon combustion (or in a fuel cell). By examining the heats of combustion of various fuels, we can see exactly how much energy it would take to create those fuels (conservation of energy). To split water requires 286 kJ/mol; when done by electrolysis, it requires 1.23 V (at a minimum). Recombining the hydrogen and oxygen generates the same 286 kJ/mol or produces 1.23 V in a fuel cell (under ideal conditions). Hydrogen is a very intense source of energy. Since a mole of hydrogen only weighs 2 g, as a fuel, its energy content is 143 kJ/g or 34 kcal/g. Compared to other fuels, hydrogen is incredibly energy rich. This is why you often hear speculation about a hydrogen economy (as opposed to a fossil fuel economy). Be sure to remember, however, that hydrogen is just the fuel (i.e., a carrier of energy, not the energy source itself). The amount of free hydrogen that exists in nature is negligible—it's much more stable as water. So when someone is touting hydrogen as the fuel of the future, be sure to ask where they plan to get the hydrogen. It would have to be from water, using solar energy, of course.

Energy content of fuels (or the energy required to create the fuels) in kilojoules per gram:

Hydrogen:

$$H_2 + 0.5\,O_2 \rightarrow H_2O + 286 \text{ kJ/mol } (143 \text{ kJ/g of hydrogen})$$

Natural gas:

$$CH_4 + 2\,O_2 \rightarrow CO_2 + 2\,H_2O + 802 \text{ kJ/mol } (50 \text{ kJ/g of methane})$$

Gasoline:

$$C_8H_{18} + 12.5\,O_2 \rightarrow 8\,CO_2 + 9\,H_2O + 5244 \text{ kJ/mol (46 kJ/g of octane)}$$

Alcohol:

$$C_2H_5OH + 3\,O_2 \rightarrow 2\,CO_2 + 3\,H_2O + 1380 \text{ kJ/mol (30 kJ/g of ethanol)}$$

Wood:

$$(CH_2O) + O_2 \rightarrow CO_2 + H_2O + 470 \text{ kJ/mol (16 kJ/g of carbohydrate)}$$

The last reaction is the exact reverse of photosynthesis. Since reversing any of the reactions above is so difficult to accomplish chemically, many scientists are simply letting nature do the work with natural photosynthesis and attempting to adapt or modify the process. For example, the photosynthetic process has been diverted by adding appropriate enzymes (hydrogenases) that take the early products of photosynthesis and use them to reduce water to generate hydrogen. Other examples work with the natural products of photosynthesis and modify them afterward, for example, by converting carbohydrates to ethanol (fermentation). Others are seeking novel plants whose photosynthetic products are useful directly. For example, in Brazil, the "diesel tree" oozes a latex that is usable as diesel fuel with minimal processing. Using natural photosynthesis to drive the production of biofuels is a growing area.

## TO CONCLUDE

I hope I've been able to give you an idea of the grand challenges and grand excitement that is chemistry these days. It's a way of thinking about the world, right down to the level of molecules. I always tell my students I want them to "think like molecules"—what do molecules look like, what do they want to do, and why do they want to do that? If you can think like a molecule, then you might be able to convince molecules to do what you want them to do. Then you're a chemist.

I'm getting old enough that people are starting to ask me whether I'm getting ready to retire soon. I always answer the same way—"not as long as I'm having fun and doing something important." Both those points remain very true for me. I wish the same for you. Have a great time with your chemistry courses and keep your mind open to all the possibilities ahead of you. You've always been a great kid, and I'm looking forward to hearing all about what you choose to do with your life.

Love,

Uncle Carl

## FURTHER READING

Bard, A. J.; Fox, M. A. Artificial photosynthesis: Solar splitting of water to hydrogen and oxygen. *Accounts of Chemical Research* 1995, *28*, 141–145.

Hambourger, M.; Moore, G. F.; Kramer, D. M.; Gust, D.; Moore, A. L.; Moore, T. A. Biology and technology for photochemical fuel production. *Chemical Society Reviews* 2009, *38*, 25–35.

Lewis, N. S. Powering the planet. *MRS Bulletin* 2007, *32*, 808–820.

O'Regan, B.; Grätzel, M. A low-cost, high-efficiency solar cell based on dye-sensitized colloidal $TiO_2$ films. *Nature* 1991, *353*, 737–740.

Sarquis, M.; Sarquis, J. *Fun with Chemistry*, Vols. 1–2, Institute for Chemical Education, Madison, WI, 1995.

Shakashiri, B. Z. *Chemical Demonstrations: A Handbook for Teachers of Chemistry*, Vols. 1–4, University of Wisconsin, Madison, WI, 1983–1992.

# 17

# Metals, Microbes, and Solar Fuel

**Harry B. Gray**
*California Institute of Technology*

**John S. Magyar**
*Barnard College*

Harry B. Gray is the Arnold O. Beckman professor of chemistry and founding director of the Beckman Institute at the California Institute of Technology in Pasadena, California. After graduate work at Northwestern University and postdoctoral research at the University of Copenhagen, he joined the chemistry faculty at Columbia University, where in the early 1960s he developed ligand field theory to interpret the electronic structures and substitution reactions of metal complexes. After moving to Caltech in 1966, he began work in biological inorganic chemistry and inorganic photochemistry, where, working with Ru-modified proteins in the early 1980s, he and coworkers discovered that electrons can tunnel rapidly over long molecular distances through folded polypeptide structures. In the years following, he and J. R. Winkler developed laser flash–quench methods that opened the way for elucidation of the factors that control electron flow through proteins that are essential components of respiratory, photosynthetic, and other biological oxidation–reduction systems.

*Letters to a Young Chemist*, First Edition. Edited by Abhik Ghosh.
© 2011 John Wiley & Sons, Inc. Published 2011 by John Wiley & Sons, Inc.

John S. Magyar is assistant professor of chemistry at Barnard College, the liberal arts college for women in New York affiliated with Columbia University. He earned his AB with honors in chemistry from Dartmouth College, in Hanover, New Hampshire, where he did undergraduate research with Dean E. Wilcox on the spectroscopy of nitric oxide binding to cobalamins. He earned his PhD with Hilary Arnold Godwin at Northwestern University, Evanston, Illinois, investigating molecular mechanisms of lead poisoning. He was a postdoctoral scholar in chemistry with Harry Gray in the Beckman Institute at Caltech, Pasadena, California, where he studied protein dynamics and electron transfer kinetics in cytochrome *c*. At Barnard, Magyar and his undergraduate research students are working to elucidate mechanisms of metal uptake and regulation in environmentally important microorganisms, including the marine cyanobacterium *Prochlorococcus*.

Dear Angela,

It was delightful to see you last weekend! We are so glad you were able to visit us in Pasadena while you were in Los Angeles to see the Padres at Dodger Stadium. (We hear it was quite an exciting game!) We hope your trip back to the University of California, San Diego (UCSD) was uneventful, and that the traffic on Interstate 5 was not too bad. (All those cars, burning gasoline, and spewing carbon dioxide into the atmosphere ... there has to be a better way!)

We've been thinking a lot this week about the questions you asked us Saturday evening. You expressed dismay about the current state of the environment and of the changing global climate, and concern about human dependence on fossil energy. You asked, what can a young person do?

Our answer: chemistry! Allow us to explain.

To put yourself in the proper frame of mind, take a walk down the hill from the main UCSD campus in La Jolla, down to the shore. Walk out on the Scripps Pier, or along the beach looking out past the surfers and seals toward the horizon. What you see is a vast expanse of water, water that covers 71% of Earth's surface. The single most apparent feature of the Earth from space is the global ocean. It holds 97% of the Earth's water, and 95% of it is unexplored.

Remarkably, one way to learn more about life in the oceans is from space. From high above the Earth, NASA imaging satellites can

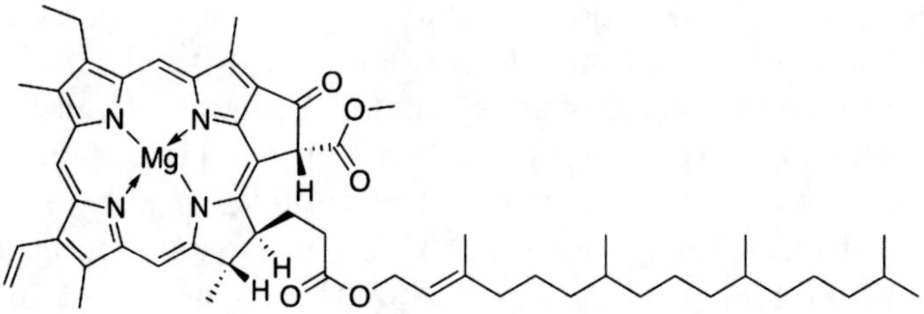

**Figure 17.1.** Chlorophyll a, the green pigment in plants, contains magnesium.

detect the characteristic green absorption, and red orange fluorescence, of chlorophyll. Chlorophyll is the molecule that gives green
plants their color, and it is used by the plant to trap energy from
sunlight and to transfer electrons through the cell (Figure 17.1). By
looking at ocean color, scientists using these satellites can track
blooms of photosynthetic plankton across the globe, and they can
relate the occurrence of these blooms to environmental factors and
conditions.

Now, from the pier or the beach, look closely at the water at your
feet. The water may appear clear, but it is home to millions of microorganisms. In surface waters, many of these microorganisms are photosynthetic, using sunlight as their primary energy source. In fact,
marine phytoplankton account for approximately 50% of all photosynthesis on Earth. Amazingly, even though they play such a critical
role in the global carbon cycle, we still don't know who all the players
are in this microbial community—and we certainly don't know all the
details about how they live.

One of the most abundant marine phytoplankton is a tiny microbe
called *Prochlorococcus*. Twenty years ago, no one knew *Prochlorococcus*
even existed, until it was discovered by Penny Chisholm and her
team at MIT. Now, we know that *Prochlorococcus* is responsible for
about one-fifth of global photosynthesis. Simply amazing! With the
modern tools of oceanography, genomics, and analytical chemistry,
we're learning a lot about what *Prochlorococcus* and other marine
phytoplankton need to grow and to thrive. One of those key requirements is acquisition of transition metals, such as iron and cobalt.

## METALS, MARINE MICROBES, AND CLIMATE

Concentrations of most transition metals in the surface oceans are
remarkably low, so figuring out how microorganisms acquire the

metals they need is an important challenge for bioinorganic chemists. Much of the work to date has focused, not surprisingly, on iron. Iron is an essential component of many metalloproteins. It's used in electron transfer and in respiration. It's a key component of hydrogenases, the proteins that catalyze the splitting of dihydrogen into protons and electrons (more on that later). Iron is important for a wide range of biological processes, but there's not a lot of iron in most parts of the surface oceans. Iron concentrations are in the nanomolar range; compare that to micromolar to millimolar concentrations of nitrogen and carbon. In fact, iron is a limiting nutrient in many areas of the ocean. Add iron and more phytoplankton grow. Just like green plants on land, phytoplankton use the energy from sunlight to convert carbon dioxide and water into oxygen and carbohydrates, which the plant can then use (in respiration) as fuel.

It has been suggested that one way to sequester the large amounts of carbon dioxide produced from fossil fuel combustion would be to fertilize the oceans with iron, thus causing a bloom of phytoplankton. In the ideal scenario for long-term carbon sequestration, these phytoplankton would then sink to the bottom of the ocean, removing the $CO_2$ they had taken up from the active carbon cycle. Large-scale experiments have even been done, involving multiple oceanographic research ships and dozens of scientists, to test this scheme at sea. It turns out that dumping tons of iron into the ocean does cause a phytoplankton bloom, a bloom large enough to be seen from space by the chlorophyll-observing satellites. But in terms of carbon sequestration, this plan is a bust. The phytoplankton bloom is short-lived, lasting only until the iron is used up. Moreover, the phytoplankton aren't removed from the carbon cycle. They are immediately grazed by zooplankton and larger animals, and the very temporarily sequestered carbon is returned to the atmosphere.

What this study does emphasize, though, is (1) how important trace metals such as iron are in the biology of the marine environment and (2) how little we know about the bioinorganic chemistry, as well as the ecology, of these environments. There's a lot for a young chemist to tackle!

It was a study by a chemist at UCSD, Angela, that provided the data showing that atmospheric carbon dioxide levels are rising. In 1958, Charles David Keeling, a scientist at the Scripps Institution of Oceanography at UCSD, started measuring carbon dioxide concentrations in the air above the South Pole and above Mauna Loa, Hawaii. Budget cuts forced him to discontinue the Antarctic measurements after a few years, but the measurements at Mauna Loa have continued to the present day. These data clearly show that the atmospheric $CO_2$ concentration is rising, year after year. Comparison with data from air bubbles trapped in ice cores shows that not only are $CO_2$ levels going up fast but they are also much higher than they've been anytime in the last 800,000 years, and that atmospheric $CO_2$ concentration and temperature are closely coupled. Burning fossil fuels has had a tremendous effect on the global environment.

Keeling earned a PhD in physical chemistry from Northwestern University and then was a postdoctoral scholar at Caltech in geological sciences. At Caltech, and later at Scripps, he was able to combine his background in chemistry with his love of the outdoors, devising instruments to make these remarkably important environmental chemical measurements. Like Keeling, generations of analytical chemists and oceanographers have worked hard to develop ingenious methods for determining marine metal concentrations. More and more good data are being collected. Now is an exciting time, Angela, to be a marine bioinorganic chemist, for it is now possible to answer questions that previously were unanswerable. It is possible to measure metal concentrations at sea and to relate those levels to the microbial communities observed there. Using observations made by oceanographers in the field and in the lab, a bioinorganic chemist can now also begin to probe molecular mechanisms of metal ion uptake.

That's the kind of work that's happening in the Magyar lab at Barnard College. We've used the genome sequence of *Prochlorococcus* to identify proteins that we predict are involved in regulating metal uptake or in transporting metals, such as cobalt. A team of talented undergraduates is using standard molecular biology techniques to make large quantities of these proteins, working out ways to purify them, and using a variety of analytical and spectroscopic techniques to study metal binding. In collaboration with Mak Saito and his group at the Woods Hole Oceanographic Institution, we can relate our study of individual metal–protein interactions to studies of *Prochlorococcus* grown in the lab under different environmental conditions and to field studies of metals and microbes at sea.

Studying the bioinorganic chemistry of the oceans can help us understand effects of climate change on ecosystems and may even help us figure out something to do with all our extra carbon dioxide. There are a lot of fundamental discoveries still to be made, and new tools of genomics and molecular microbiology open up many exciting chemical problems to study!

## Energy

Another aspect of the climate problem that chemists think about, Angela, is how we can replace fossil fuels as the world's primary energy source.

One possibility is nuclear power. As you drove down Interstate 5 last weekend, returning to San Diego from Los Angeles, you passed the San Onofre nuclear power plant. The two San Onofre generators each produce 1.1 billion watts of power, 1.1 GW, enough to provide electricity for 2.8 million households. In 2001, the world used approximately 14 *trillion* watts, 14 TW. Our friend at Caltech, Nate Lewis, has pointed out that stabilizing atmospheric $CO_2$ levels will require the production of more than 10 TW of *additional* carbon-neutral energy between now and 2050. To use nuclear fission to generate an additional 10 TW, you would need to build a new nuclear reactor the size of one of the San Onofre generators every other day for the next 50 years! Nuclear power is likely an important part of a comprehensive energy strategy, Angela, but it alone cannot solve our energy woes. Neither can wind, nor hydropower, nor biofuels, although all are worth investigating. The scale of the problem is simply too large.

The only resource we have that can, in principle, meet all our needs is solar energy. More energy from sunlight hits the Earth in an hour than the entire world uses in a year! If we can find a way to harness that energy—cheaply, efficiently, and on a very large scale—we can solve our growing energy crisis. We hope it will help ameliorate the climate crisis, too.

We know how to make electricity from sunlight. We can do that very effectively with silicon solar panels, larger versions of the solar cell on your calculator. Chemists (including your uncle at Portland State University, Carl Wamser) and other scientists have been working hard to develop new kinds of dye-sensitized solar cells for electricity production, which could be even cheaper to produce and to use.

These are important efforts, but alone, they are not enough. One of the problems with solar electricity production, Angela, is that the places that are geographically best for collecting energy from the sun (or the winds, for that matter) are far from population centers. Collecting sunlight and generating electricity in the middle of Wyoming may generate lots of electricity, but there is no one there to use it! Building transmission lines from Wyoming or Arizona to transport that electricity to the major urban centers would be very expensive. For the large-scale replacement of fossil energy, it will be essential to turn sunlight into chemical fuels, not just into electricity.

These fuels can then be transported wherever they are needed and used whether the sun is shining or not!

After all, this is what nature does already! Think back to the phytoplankton in the waters off the Scripps Pier ... or to your houseplants, or the trees in your backyard. The green pigments (chlorophyll) in the leaves of these plants, or in the phytoplankton, collect energy from sunlight and transfer that energy to a large assemblage of proteins known as photosystem II (Figure 17.2). In photosystem II, the oxygen-evolving complex catalyzes the reaction:

$$2H_2O \rightarrow O_2 + 4H^+ + 4\,e^-.$$

A

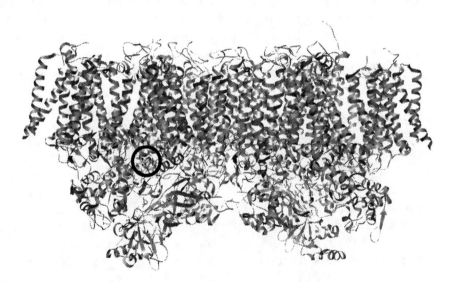

B

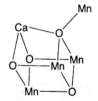

**Figure 17.2.** (a) The structure of photosystem II at 3.5-Å resolution, as determined by X-ray crystallography. Source: Ferreira, K. N.; Iverson, T. M.; Maghlaoui, K.; Barber, J.; Iwata, S. Architecture of the Photosynthetic Oxygen-Evolving Center. Science 2004, 303, 1831–1838, PDB: 1S5L). The oxygen-evolving complex is circled in black. (b) A closer view of the oxygen-evolving complex, which is made up of manganese, calcium, and oxygen atoms arranged in a nearly cubic geometry.

Plants then use the protons and electrons that are produced in this reaction to transform carbon dioxide into carbohydrates. In other words, the plants are converting energy from sunlight into chemical fuel.

The chemist's goal is similar: to use sunlight to split water into hydrogen and oxygen to make fuel. Unlike the plants, however, we would prefer to make dioxygen and dihydrogen as our reaction products instead of protons:

$$2H_2O \rightarrow O_2 + 2H_2.$$

Not only do we not really want the complications of carbohydrate synthesis in our fuel production but we also would prefer to end up with a fuel that will not produce $CO_2$ when it is burned. Burning $H_2$ in air to release energy is simply the reverse of the water-splitting reaction, so the only by-product is clean water. (In practice, the dihydrogen that we produce in a water-splitting reaction would more likely be oxidized in a fuel cell than literally burned; but either way, water is the only waste product!)

So, what are the challenges? Well, although we now know an awful lot about how photosynthesis works, there's still much we don't understand about the molecular mechanisms of biological water splitting. The other problem with biological water splitting is that both leaves and plankton have very short life cycles! The photosynthetic apparatus is destroyed and rebuilt many times a day ... that's certainly not practical for a solar cell for your car or your house! Another issue is cost and availability of raw materials. If we are to replace fossil energy with solar fuels, the scale we must work on is enormous! Solar cells made of materials that are too rare or too expensive will not solve the global problem, no matter how elegantly they may be designed or how well they might work. Successful solar water-splitting cells must be made of cheap, readily available materials so that they can be produced and used on massive scales.

## BIOLOGICAL INSPIRATIONS AND INORGANIC CHEMISTRY

The way nature splits water is a good place to start. The part of a green leaf where photosynthesis happens is called the chloroplast. These chloroplasts look a lot like cyanobacteria, such as *Prochlorococcus*, and they do they same job: converting energy from sunlight into fuel for the cell. (There's some evidence that chloroplasts evolved directly from cyanobacteria.)

Back in 1985, Eli Greenbaum at the Oak Ridge National Laboratory in Tennessee coated chloroplasts with platinum. It's been known for more than 200 years that platinum can be used to split water into hydrogen and oxygen. In 1800, the British chemists William Nicholson and Anthony Carlisle reported splitting water into hydrogen and oxygen by electrolysis, using the brand-new invention of Alessandro Volta, the galvanic pile. Using his platinized chloroplasts, Greenbaum was able to split water with the energy from sunlight, without converting the solar energy to electricity first. A problem with Greenbaum's system is the biological component: the chloroplasts degrade very quickly under the harsh conditions of these reactions. Leaves, as we mentioned before, have a self-repair feature built in. Separated from a growing plant, chloroplasts cannot repair themselves and turn into sludge pretty quickly.

It's possible to make a really efficient water-splitting cell that doesn't use any of the parts of a leaf. John Turner and the chemists and engineers at National Renewable Energy Laboratory (NREL) have done just that. Their cell uses the semiconductors gallium arsenide (which absorbs reddish light) and gallium indium phosphide (which absorbs bluish light), and with these two materials, they can use most of the solar energy that hits the cell. Not much goes to waste. The water is split with a platinum catalyst. It works beautifully, and it's terrifically efficient!

There are three problems. First, the NREL cell is too expensive—it costs more than $10,000 per square centimeter to make. Second, it uses platinum—there isn't enough platinum on Earth to work on the scale we need, also a problem for the Greenbaum and Nicholson–Carlisle systems. Finally, the NREL cell is too environmentally unfriendly; arsenic is highly toxic. We must do better!

What chemists have learned from these studies and many others is to turn their attention to the elements that nature uses, elements like manganese, iron, nickel, and cobalt. The oxygen-evolving complex in photosynthesis is a cluster of manganese (and oxygen and calcium) atoms. All the known hydrogenase proteins, mentioned earlier, contain iron; some also contain nickel (Figure 17.3). These metals are Earth-abundant and are much more environmentally benign than arsenic or platinum.

Developing new, robust, solar-powered water-splitting catalytic systems using Earth-abundant materials is and will continue to be

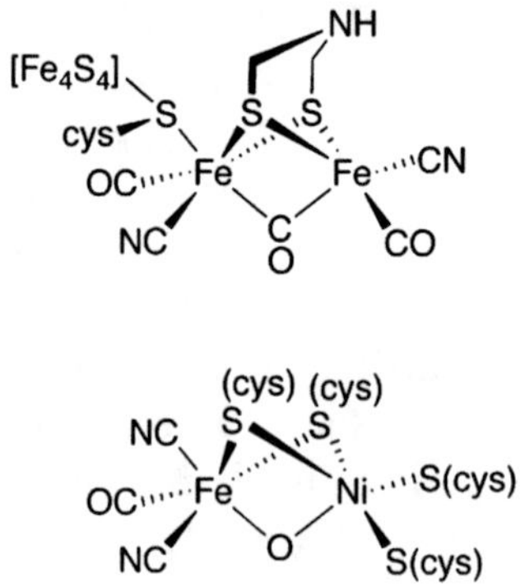

**Figure 17.3.** The active sites in all known hydrogenase proteins contain iron; many also contain nickel.

one of the grand challenges of twenty-first century chemistry. There is much to do!

## Powering the Planet

We are glad that you had a chance to see some of the frontline work in this area on your recent visit to Caltech. The Center for Chemical Innovation (CCI) "Powering the Planet" solar energy research program involving Caltech and many other institutions is going strong, and lots of "CCI Solar" people are working hard on fundamental science aimed at developing a nanoscale water-splitting device. Any water-splitting device needs some way to collect sunlight. It also needs two catalytic parts, one to produce hydrogen and one to produce oxygen. These catalytic parts need to be separated from each other in space, probably by a membrane as they are in chloroplasts, so that the hydrogen and oxygen don't recombine explosively! Nate Lewis's group at Caltech is working hard to develop new materials for the membrane between the two catalytic sides of the cell and the light-collecting apparatus to go with it. Silicon nanorods will be embedded in the membrane to collect sunlight. The nanorods will stick out from both sides of the membrane, and they'll be coated with two different catalysts. One side of the membrane will have blue light-absorbing nanorods to interact with an $O_2$-evolving catalyst. When water is split to evolve oxygen, protons and electrons are generated. The other side of the membrane will absorb red light, and a second catalyst will combine these protons and electrons to produce dihydrogen.

## Making Hydrogen

A Caltech grad student, Jillian Dempsey, is making wonderful progress in this area. As we discussed earlier, we want to use Earth-abundant materials; and we don't want biological materials like proteins or chloroplasts around to decompose. Simply mimicking the metal centers in protein active sites doesn't work. For activity, biological systems depend on having not just the metals but also all the rest of the protein around.

The molecule that Jillian is studying was originally developed by Jim Espenson at Iowa State in the 1980s and was improved by Xile Hu when he was a postdoc with Jonas Peters at Caltech. (Xile is now an assistant professor at the École Polytechnique Fédérale de Lausanne, Switzerland.) Jillian's working very hard to elucidate the mechanism of dihydrogen production by this molecule, which features a cobalt atom as its center. She's learning through laser flash–quench experiments (pioneered by Caltech's Jay Winkler) whether it's necessary to have two cobalt–hydride complexes colliding to release $H_2$, or whether a single proton can remove a hydride from a single cobalt–hydride complex. Do the cobalt atoms work alone or do they have to work in pairs? If they work in pairs, then a tethered binuclear cobalt system like the one synthesized by Caltech undergraduate Carolyn Valdez would help to increase the rate of reaction. (Perhaps you and Carolyn will have a chance to meet the next time you visit, and she can tell you more about her progress!)

## Making Oxygen

Even more challenging than the hydrogen-evolving side of the cell is the oxygen-evolving catalyst, which MIT chemists Dan Nocera and Kit Cummins are working on. One of the challenges is that four electrons and four protons must be removed from two water molecules to make dioxygen. This is a very tall order! Nature does this reaction using multiple metals—the oxygen-evolving complex in photosystem II has four manganese atoms and a calcium atom. Removing four electrons from two water molecules using four metal ions (one electron per metal) is much easier than extracting all four from a single metal!

The good news is that progress has been made in the CCI Solar program. Dan Nocera's group has recently reported a cobalt-based inorganic catalyst that is very promising, and his team at MIT is now working hard to figure out just how it functions. And at Caltech, grad student Kyle Lancaster is taking a biomimetic approach, starting with an enzyme that is known to catalyze the four-electron reduction of $O_2$. Since catalysts work to accelerate the rates of both forward and reverse reactions, Kyle thinks it should be possible through some tweaking of this enzyme system to evolve oxygen instead of consuming it. The challenge is to modify the enzyme so it will operate at a redox potential that is favorable for water oxidation. At the same time, Kyle will have to find derivatives that minimize protein degradation. What is certain is that in the process, he (and all of us) will learn a lot about the fundamentals of multielectron oxidation–reduction processes. Understanding these fundamentals will allow the development of even better oxygen-evolving catalysts down the road.

Angela, I hope we've been able to give you a flavor of the kinds of exciting chemical research projects that are possible in areas related to energy and climate. There is much to learn, and you can be sure that fundamental research in these areas will make enormous contributions to the health and well-being of our planet. I hope you will consider joining us in this effort!

Best wishes,

John and Harry

## FURTHER READING

Bertini, I.; Gray, H. B.; Stiefel, E. I.; Valentine, J. S. *Biological Inorganic Chemistry: Structure and Reactivity*, University Science Books, Sausalito, CA, 2007.

Butler, A. Acquisition and utilization of transition metal ions by marine organisms. *Science* 1998, *281*, 207–210.

Chisholm, S. W. The cells that rule the seas. *Scientific American* 2003, 52–53.

Goodstein, D. L. *Out of Gas: The End of the Age of Oil*, W. W. Norton, New York, 2004.

Gray, H. B. Powering the planet with solar fuel. *Nature Chemistry* 2009, *1*, 7.

Morel, F. M. M.; Price, N. G. The biogeochemical cycles of trace metals in the oceans. *Science* 2003, *300*, 944–947.

# Index

Page numbers in *italics* refer to Figures; those in **bold** to Tables.

*Letters to a Young Chemist*, First Edition. Edited by Abhik Ghosh.
© 2011 John Wiley & Sons, Inc. Published 2011 by John Wiley & Sons, Inc.